Practical Node-RED Programming

Learn powerful visual programming techniques and best practices for the web and IoT

Taiji Hagino

BIRMINGHAM—MUMBAI

Practical Node-RED Programming

Associate Group Product Manager: Pavan Ramchandani

Publishing Product Manager: Kaustubh Manglurkar

Senior Editor: Sofi Rogers

Content Development Editor: Rakhi Patel

Technical Editor: Saurabh Kadave

Copy Editor: Safis Editing

Language Support Editor: Safis Editing

Project Coordinator: Divij Kotian

Proofreader: Safis Editing

Indexer: Manju Arasan

Production Designer: Alishon Mendonca

First published: March 2021

Production reference: 1190321

Published by Packt Publishing Ltd.

Livery Place

35 Livery Street

Birmingham

B3 2PB, UK.

ISBN 978-1-80020-159-0

www.packt.com

To my colleague, Nick O'Leary, and Node-RED Community co-organizers, Atsushi Kojo, Seigo Tanaka, and Kazuhito Yokoi, I would like to thank you for taking time from your busy schedules to help me with the book. I would also like to thank my wife, Akiko, for being my loving partner and supporting me throughout writing this book and always.

– Taiji Hagino

Foreword

Taiji has been deeply involved with the Node-RED User Group Japan since its creation. In his developer advocate role, he has worked with many users to help them build meaningful applications with Node-RED. This book reflects Taiji's skills and experience with the project and will be a great resource for many readers.

This book will provide you with a good introduction to Node-RED and give you a sense of how quickly you can get started with creating applications. The examples in each chapter will give you a taste of how much can be achieved with very little coding.

I hope it inspires you to continue building with Node-RED and to explore everything that is possible.

Nick O'Leary

Co-creator of Node-RED

Taiji has extensive development knowledge in the web/cloud, mobile, IoT, blockchain, and so on. We have known each other since the inception of the Node-RED User Group Japan 5 years ago.

Taiji has been an active contributor to the Node-RED community since the early days of Node-RED, running Node-RED meetups with us. He was a co-author of the book *First Node-RED* published by the Node-RED User Group Japan 3 years ago.

For more than 5 years, Node-RED has been evolving to meet the needs of developers around the world. During this time, Taiji has been a key member of IBM and has been active in Developer Advocates and Developer Relations.

In addition, Taiji has been able to gain a deep understanding of other languages and cultures through his global activities as a developer advocate and in developer relations.

Taiji has used his knowledge and experience from these global activities to organize the Node-RED Conference Tokyo, a global Node-RED event that has run for two consecutive years, where he has used his global skills to communicate with speakers from overseas and to facilitate the day of the event.

I believe Taiji will continue to serve as a global career model for Japanese developers and will be a key player in the development of the Node-RED community around the world.

Atsushi Kojo

Chief research officer at Uhuru Corporation

Taiji and I have been working together at the Node-RED User Group Japan for 5 years. He is one of the user group organizers. Taiji is especially looking globally with the aim of sharing technological possibilities, such as setting up a meeting between the organizer of a Japanese user group and the Node-RED development team at IBM Hursley. Recently, we held Node-RED Con Tokyo 2019 and 2020 together. Taiji has also carried out an important role as an online moderator and manager.

Taiji has written various blogs where he has shared his immense knowledge of Node-RED smartly. The source of his knowledge comes from his great experience as an excellent developer and developer advocate at IBM. He has gained experience with business use cases and development knowledge such as IoT, mobile applications, cloud technologies, databases, and blockchain in his developer relations activity.

He has a strong understanding of the synergies and difficulties of combining each technology. Many developers find Node-RED attractive because of him. This book represents how knowledgeable he is as a developer.

Read this book and discover how wonderful it is to combine various technologies such as IoT and the cloud using Node-RED, and expand your possibilities as a developer.

Seigo Tanaka

President, 1ft seabass

Contributors

About the author

After becoming a software engineer, **Taiji Hagino** started Accurate System Ltd. with his amazing software development experience. After working as a system integrator of a subsidiary of a general trading company, he now works as a developer advocate in the IBM Global team, developing DevRel (developer relations), a marketing approach to engineers. He also works as a lecturer at the Faculty of Informatics, University of Tsukuba. Works he has authored include *Developer Marketing DevRel Q&A* (Impress R&D), *First Node-RED*, *Practical Node-RED Application Manual* (Kogakusha), and so on. He has been awarded Microsoft MVP and was previously a musician and a hairdresser.

I want to thank all the people who have been close to me and supported me throughout writing this book, especially my wife, Akiko, and my family.

About the reviewers

Nick O'Leary is an open source developer and leads the OpenJS Node-RED project. He spends his time playing with IoT technologies, having worked on projects ranging from smart meter energy monitoring to retrofitting sensors to industrial manufacturing lines with Raspberry Pis and Arduinos. With a background in pervasive messaging, he is a contributor to the Eclipse Paho project and sits on the OASIS MQTT Technical Committee and the OpenJS Cross Project Council.

Kazuhito Yokoi works for OSS Solution Center in Hitachi, Ltd. as a software engineer. On GitHub, he is a member of the Node-RED project. Hitachi has used Node-RED in their IoT platform, Lumada. To improve the code quality and add new features, his team joined the Node-RED project as contributors. For 4 years, 19 contributors in his team have added over 700 commits and 80,000 lines to the project. Currently, they are contributing to not only Node-RED but also sub-projects such as a node generator to generate nodes from various sources without coding, and a Node-RED installer to set up Node-RED without CLI operations. He held sessions about Node-RED at the Open Source Summit Japan 2020, Node+JS Interactive 2018, and other global conferences.

Table of Contents

3

Understanding Node-RED Characteristics by Creating Basic Flows

4

Learning the Major Nodes

Section 2: Mastering Node-RED

5

Implementing Node-RED Locally

6

Implementing Node-RED in the Cloud

7

Calling a Web API from Node-RED

8

Using the Project Feature with Git

Section 3: Practical Matters

9

Creating a ToDo Application with Node-RED

10

Handling Sensor Data on the Raspberry Pi

11

Visualize Data by Creating a Server-Side Application in the IBM Cloud

12

Developing a Chatbot Application Using Slack and IBM Watson

13

Creating and Publishing Your Own Node on the Node-RED Library

Appendix

Node-RED User Community

Other Books You May Enjoy

Index

Preface

Node-RED is a flow-based programming tool that was made by Node.js. This tool is mainly used for connecting IoT devices and software applications. However, it can cover not only IoT but also standard web applications.

Node-RED is expanding as a no-code/low-code programming tool. This book covers the basics of how to use it, including new features that have been released from version 1.2, as well as advanced tutorials.

Who this book is for

This book is best for those who are learning about software programming for the first time with no-code/low-code programming tools. Node-RED is a flow-based programming tool, and this tool can build web applications for any software applications easily, such as IoT data handling, standard web applications, web APIs, and so on. So, this book will help web application developers and IoT engineers.

What this book covers

Chapter 1, *Introducing Node-RED and Flow-Based Programming*, teaches us what Node-RED is. The content also touches on flow-based programming, explaining why Node-RED was developed and what it is used for. Understanding this new tool, Node-RED, is helpful to improve our programming experience.

Chapter 2, *Setting Up the Development Environment*, covers setting up the development environment by installing Node-RED. Node-RED can be installed for any OS Node.js can run, such as Windows, macOS, Rasberry Pi OS, and so on. We install Node-RED on each environment with the command line or using the installer. This chapter covers important notes for specific OSes.

Chapter 3, *Understanding Node-RED Characteristics by Creating Basic Flows*, teaches us about the basic usage of Node-RED. In Node-RED, various functions are used with parts called nodes. In Node-RED, we create an application with a concept called a flow, like a workflow. We will create a sample flow by combining basic nodes.

Chapter 4, Learning the Major Nodes, teaches us how to utilize more nodes. We will not only learn about the nodes provided by Node-RED by default but also how to acquire various nodes published on the internet by the community and how to use them.

Chapter 5, Implementing Node-RED Locally, teaches us best practices for leveraging Node-RED in our local environment, our desktop environment. Since Node-RED is a tool based on Node.js, it is good at building server-side applications. However, servers aren't just on beyond the network. It is possible to use it more conveniently by using Node-RED in a virtual runtime on the local environment of an edge device such as Raspberry Pi.

Chapter 6, Implementing Node-RED in the Cloud, teaches us best practices for leveraging Node-RED on a cloud platform. Since Node-RED is a tool based on Node.js, it is good at building server-side applications. It is possible to use it more conveniently by using Node-RED on any cloud platform, so we will make flows with Node-RED on IBM Cloud as one of the use cases with cloud platforms.

Chapter 7, Calling a Web API from Node-RED, teaches us how to utilize the web API from Node-RED. In order to maximize the appeal of web applications, it is essential to link with various web APIs. Its application development architecture is no exception in Node-RED. Understanding the difference between calling a web API from a regular Node.js application and calling it from Node-RED can help us get the most out of Node-RED.

Chapter 8, Using the Project Feature with Git, teaches us how to use source code version control tools in Node-RED. With Node-RED, the project function is available in version 1.x and later. The project function can be linked with each source code version control tool based on Git. By versioning the flows with the repository, our development will be accelerated.

Chapter 9, Creating a ToDo Application with Node-RED, teaches us how to develop standard web applications with Node-RED. The web application here is a simple ToDo application. The architecture of the entire application is very simple and will help us understand how to develop a web application, including the user interface, using Node-RED.

Chapter 10, Handling Sensor Data on the Raspberry Pi, teaches us application development methods for IoT data processing using Node-RED. Node-RED was originally developed to handle IoT data. Therefore, many of the use cases where Node-RED is still used today are IoT data processing. Node-RED passes the data acquired from sensors for each process we want to do and publishes it.

Chapter 11, Visualize Data by Creating a Server-Side Application in the IBM Cloud, teaches us about application development methods for IoT data processing using Node-RED on the cloud platform side. We usually use the data from edge devices on any cloud platform for analyzing, visualization, and so on. Node-RED handles the data subscribed from the MQTT broker and visualizes it for any purpose.

Chapter 12, Developing a Chatbot Application Using Slack and IBM Watson, teaches us how to create a chatbot application. At first glance, Node-RED and chatbots don't seem to be related, but many chatbot applications use Node-RED behind the scenes. The reason is that Node-RED can perform server-side processing on a data-by-data basis like a workflow. Here, we create a chatbot that runs on Slack, which is used worldwide.

Chapter 13, Creating and Publishing Your Own Node on the Node-RED Library, teaches us how to develop nodes ourselves. For many use cases, we can find the node for the processing we need from the Node-RED Library. This is because many nodes are exposed on the internet thanks to the contributions of many developers. Let's aid a large number of other Node-RED users by developing our own node and publishing it to the Node-RED Library.

To get the most out of this book

You will need Node-RED version 1.2 or later, Node.js version 12 or later, npm version 6 or later, and preferably the latest minor version installed on your computer. But this is the case when running Node-RED in a local environment. In the case of running on IBM Cloud, which is one of the tutorials in this book, it depends on the environment of the cloud platform. All code examples have been tested on macOS, Windows, and Raspberry Pi OS, but some chapters have command-line instructions based on macOS.

Software/hardware covered in the book	OS requirements
Node-RED 1.2	Windows, macOS, or Raspberry Pi OS
Node.js 12	Windows, macOS, or Raspberry Pi OS
npm 6	Windows, macOS, or Raspberry Pi OS

If you are using the digital version of this book, we advise you to type the code yourself or access the code via the GitHub repository (link available in the next section). Doing so will help you avoid any potential errors related to the copying and pasting of code.

Download the example code files

You can download the example code files for this book from GitHub at `https://github.com/PacktPublishing/-Practical-Node-RED-Programming`. In case there's an update to the code, it will be updated on the existing GitHub repository.

We also have other code bundles from our rich catalog of books and videos available at `https://github.com/PacktPublishing/`. Check them out!

Download the color images

We also provide a PDF file that has color images of the screenshots/diagrams used in this book. You can download it here: `https://static.packt-cdn.com/downloads/9781800201590_ColorImages.pdf`.

Conventions used

There are a number of text conventions used throughout this book.

`Code in text`: Indicates code words in text, database table names, folder names, filenames, file extensions, pathnames, dummy URLs, user input, and Twitter handles. Here is an example: "Let's attach a page heading to the body with the <h1> tag."

A block of code is set as follows:

```
// generate random number
var min = 1 ;
var max = 10 ;
var a = Math.floor( Math.random() * (max + 1 - min) ) + min ;

// set random number to message
msg.payload = a;

// return message
return msg;
```

Any command-line input or output is written as follows:

```
$ node --version
v12.18.1
```

```
$ npm -version
6.14.5
```

Bold: Indicates a new term, an important word, or words that you see onscreen. For example, words in menus or dialog boxes appear in the text like this. Here is an example: "After selecting the name and payment plan, click the **Select Region** button."

> **Tips or important notes**
> Appear like this.

Get in touch

Feedback from our readers is always welcome.

General feedback: If you have questions about any aspect of this book, mention the book title in the subject of your message and email us at customercare@packtpub.com.

Errata: Although we have taken every care to ensure the accuracy of our content, mistakes do happen. If you have found a mistake in this book, we would be grateful if you would report this to us. Please visit www.packtpub.com/support/errata, selecting your book, clicking on the Errata Submission Form link, and entering the details.

Piracy: If you come across any illegal copies of our works in any form on the Internet, we would be grateful if you would provide us with the location address or website name. Please contact us at copyright@packt.com with a link to the material.

If you are interested in becoming an author: If there is a topic that you have expertise in and you are interested in either writing or contributing to a book, please visit authors.packtpub.com.

Reviews

Please leave a review. Once you have read and used this book, why not leave a review on the site that you purchased it from? Potential readers can then see and use your unbiased opinion to make purchase decisions, we at Packt can understand what you think about our products, and our authors can see your feedback on their book. Thank you!

For more information about Packt, please visit packt.com.

Section 1: Node-RED Basics

In this section, readers will understand what a **flow-based programming (FBP)** tool is, including Node-RED, along with how to undertake IoT/web programming with it, and will learn how to use the Node-RED flow editor at a basic level.

In this section, we will cover the following chapters:

- *Chapter 1, Introducing Node-RED and Flow-Based Programming*
- *Chapter 2, Setting Up the Development Environment*
- *Chapter 3, Understanding Node-RED Characteristics by Creating Basic Flows*
- *Chapter 4, Learning the Major Nodes*

1
Introducing Node-RED and Flow-Based Programming

This chapter will help you grow from being a reader to being a Node-RED user. First, you'll learn about the history of **Flow-based programming (FBP)** tools, not just Node-RED. You will then gain a broad understanding of the entirety of Node-RED as a useful tool for building web applications and the **Internet of Things** (**IoT**) data handling, before learning what IoT and Node.js are in terms of Node-RED.

Providing technical content will help accelerate your software application development, but if you take a look at the history of the Node-RED tool itself, it will help you better understand why you need a FBP tool such as Node-RED. That is what we will be doing in this chapter.

More specifically, we'll be covering the following topics:

- What is FBP?
- What is Node-RED?
- Node-RED benefits
- Node-RED and IoT

Let's get started!

What is FBP?

So, what is FBP in the first place? It's the workflows you use in your work that you can easily imagine. Let's recall those workflows.

Workflows

In a normal workflow, boxes and wires indicate the process flow. It may be just one business design. Boxes represent processes. Box processing is defined by who, when, where, what, and how much. Sometimes, it's like explicitly writing out the flow of processing, such as by using swim lanes or placing writing definitions inside boxes. In any case, looking at the box should reveal what will be done.

On the other hand, let's try to summarize this business process as a document. Don't you think it will be complicated? Who will do what as they read it, even if they use some paragraphs well to put it together? When will you do it? It could be confusing:

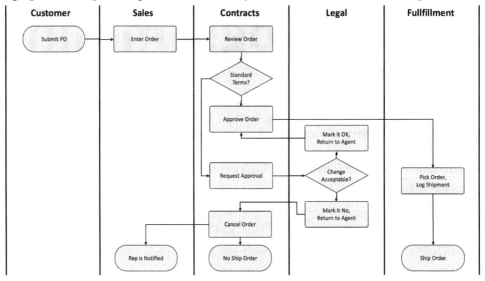

Figure 1.1 – Workflow example

Now, let's get back to software programming. FBP is a kind of concept for software programming that defines an application with a data flow. Each part of the process is there as a black box. They communicate data between connected black boxes that have been predefined. FBP is said to be component-oriented because these black-box processes can be connected repeatedly to form several applications without needing to be modified internally. Let's explore FBP in more detail.

Flow-based programming (FBP)

I think FBP is a good blend of workflow and dataflow. FBP uses a **data factory** metaphor to define an application. It sees an application as a network of asynchronous processes that start at some point and do a single sequential process that does one operation at a time until it ends, rather than communicating by using a stream of structured chunks of data. This is called an **information packet (IP)**. This view focuses on the data and its transformation process to produce the output that is needed. Networks are usually defined outside a process as a list of connections that is interpreted by a piece of software called a **scheduler**.

Processes communicate via fixed capacity connections. Connections are connected to processes using ports. The port has a specific name that is agreed on by the network definition and the process code. At this point, it is possible to execute the same code by using multiple processes. A particular IP is usually only owned by a single process or transferred between two processes. The port can be either a normal type or an array type.

FBP applications typically run faster than traditional programs, since FBP processes can continue to run as long as there is room to put in data and output to process. It does not require any special programming and makes optimal use of all the processors on the machine.

FBP has a high-level, functional style so that the behavior of the system can be easily defined; for example, in a distributed multi-party protocol such as a distributed data flow model, for accurately analyzing the criteria for determining whether a variable or statement behaves correctly:

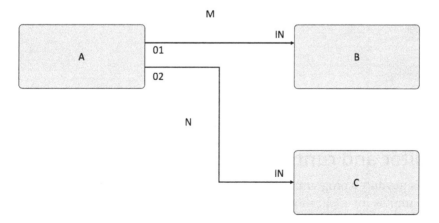

Figure 1.2 – Simple FBP design example

Now that you have a solid understanding of FBP, let's learn how Node-RED can be implemented in this way.

What is Node-RED?

Node-RED is one of the FBP tools that we have described so far. Developed by IBM's Emerging Technology Services team, Node-RED is now under the OpenJS Foundation.

Overview

FBP was invented by J. Paul Morrison in the 1970s. As we mentioned earlier, FBP describes the behavior of the application as a black box network, which in Node-RED is described as a "node." Processing is defined in each node; data is given to it, processing is performed using that data, and that data is passed to the next node. The network plays the role of allowing data to flow between the nodes.

This kind of programming method is very easy to use to make a model visually and makes it easy to access for several layer users. Anybody can understand what the flow is doing if a problem is broken down into each step. That's why you don't need to the code inside the nodes:

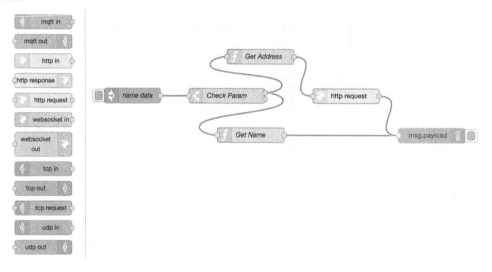

Figure 1.3 – Node-RED Flow Editor as an FBP tool

Flow editor and runtime

Node-RED is not only a programming tool but also an execution platform that wraps up the Node.js runtime for applications that are built using Node-RED.

We need to use the **flow editor** to make Node-RED applications for IoT, web services, and more. The flow editor is also a Node.js web application. We will tell you how to use flow editor clearly in *Chapter 3, Understanding Node-RED Characteristics by Creating Basic Flows*.

The flow editor, which is the core function of Node-RED, is actually a web application made with Node.js. It works with the Node.js runtime. This flow editor operates within the browser. You must select the node you want to use from the various nodes in the palette and drag it to the workspace. Wiring is the process of connecting the nodes to each other, which creates an application. The user (developer) can deploy the application to the target runtime with just one click.

The palette that contains various nodes can easily be expanded as you can install new nodes created by developers, meaning you can easily share the flow you created as a JSON file to the world. Before we explore the benefits of Node-RED, let's look at the brief history behind its creation.

History and origin of Node-RED

In early 2013, Nick-O'Leary and Dave Conway-Jones from IBM UK's Emerging Technology Services Team created Node-RED.

Originally, it was a just **proof of concept** (**PoC**) to help visualize and understand the mapping between **Message Queue Telemetry Transport** (**MQTT**) topics, but soon, it became a very popular tool that could be easily extended to various uses.

Node-RED became open source in September 2013 and remains to be developed as open source now. It became one of the founding projects of the JS Foundation in October 2016, which has since merged with the Node.js Foundation to create the OpenJS Foundation, doing so in March 2019.

The OpenJS Foundation supports the growth of JavaScript and web technologies as a neutral organization to lead and keep any projects and fund activities jointly, which is beneficial to the whole of the ecosystem. The OpenJS Foundation currently hosts over 30 open source JavaScript projects, including Appium, Dojo, jQuery, Node.js, and webpack.

Node-RED has been made available under the Apache 2 license, which makes it favorable to use in a wide range of settings, both personal and commercial:

Figure 1.4 – Dave Conway-Jones and Nick O'Leary

Why is it Called Node-RED?

The official documentation (`https://nodered.org/about/` states that the name was an easy play on words that sounded like "Code Red." This was a dead end, and Node-RED was a big improvement on what it was called in its first few days of conception. The "Node" part reflects both the flow/node programming model, as well as the underlying Node.js runtime.

Nick and Dave never did come to a conclusion on what the "RED" part stands for. "Rapid Event Developer" was one suggestion, but it's never been compelled to formalize anything. And so, the name "Node-RED" came to life.

Node-RED benefits

Let's think a little here. Why do you use cars? I think the answer is very simple and clear. First of all, we can come up with the answer that they are used as a means of transportation in a broad sense. There are other options for transportation, such as walking, bicycle, train, and bus. Then, we have the reasons for choosing a car from among these other options, as follows:

- You do not get exhausted.

- You can reach your destination quickly.

- You can move at your own pace.

- You can keep your personal space.

Of course, there are some disadvantages, but I think these are the main reasons for using a car. Although other means of transportation can also serve the same purpose, the important thing is to consider the advantages and disadvantages of each, and use the car as a transportation tool for the reason that you feel is the most suitable to you.

We can see the same situation in software. As an example, why do you use Word, Excel, and PowerPoint? You'll probably use Word because it's the most efficient way to write a document. However, you could use a word processor separately or handwrite anything. Similarly, instead of Excel, you can use any other means to make spreadsheets. There are also other means if you want to make presentation materials and make them look effective, besides PowerPoint. However, you are likely to choose the optimum tool for your situation.

Let's recall what Node-RED is for. It is a FBP tool, suitable for making data control applications for web applications and IoT. Its development environment and execution environment are browser-based applications made with Node.js, which makes their development as easy as possible.

So, what is the reason for using Node-RED, which provides these kinds of features? Do you want to avoid heavy coding? Do you not have coding skills? Yes, of course, these are also reasons to use the program.

Let's recall the example of a car. In a broad sense, our dilemma (transportation) is replaced here by developing (creating) a Node.js application for describing software tools. The transport options, such as cars, bicycles, trains, buses, ships, planes, and so on, are options, and with software development, we also have numerous options, such as using Node.js scratch, or using various frameworks of Node.js and using Node-RED. As for reasons to choose Node-RED, let's take a look at some essential points.

Simplification

When programming with Node-RED, you'll notice its simplicity. As the name no-code/low-code indicates, coding is eliminated and programming is intuitively completed with a minimal number of operations needing to be used.

Efficiency

The FBP typified by Node-RED can be completed with almost only GUI operations. Node-RED flow editor takes care of building the application execution environment, library synchronization, the **integrated development environment** (**IDE**), and editor preparation so that you can concentrate on development.

Common

As represented by object-oriented development, making the source code a common component is one of the most important ideas in development. In normal coding-based development, each common component exists in functions and classes, but in Node-RED, they exist as an easy-to-understand node (just a box). If you don't have a node as a common component you want to use, anyone can create one immediately and publish it to the world.

High quality

High quality is the true value of flow-based and visual programming. Each node provided as a component is a complete module that has been unit tested. As a result, app authors can focus on checking the operation at the join level without worrying about the contents of node. This is a big factor that eliminates human error at the single level and ensures high quality.

Open source

Node-RED is an open source piece of software. Therefore, it can be used flexibly under the Apache2 license. Some are developing their own services based on Node-RED, while others are changing to their own UI and deploying it as built-in. As we mentioned previously, we have also established a platform where we can publish our own developed node so that anyone can use it.

Node-RED library

The library indexes all Node-RED modules published to the public npm repository (https://www.npmjs.com/), assuming they follow the proper packaging guidelines.

This is the area in which we've seen the most community contribution, with well over 2,000 nodes available – which means there's something for everyone:

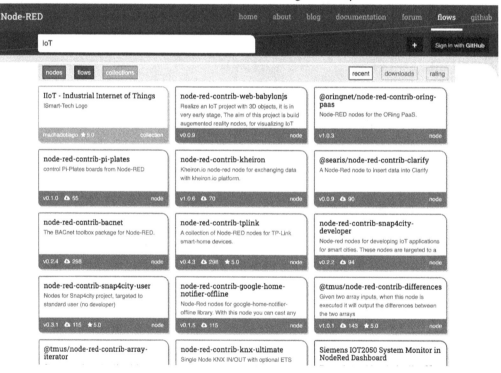

Figure 1.5 – Node-RED library

Various platforms

Node-RED can be used on various platforms. That's because Node-RED itself is a Node.js application, as we mentioned previously. If you have a runtime environment for Node.js, you can run it. It is mostly used on Edge devices, cloud services, and in embedded formats.

You can get a sense of this by understanding the relationship between Node-RED and IoT and the architecture of IoT, which will be explained in the next section.

Node-RED and IoT

Again, Node-RED is a **virtual environment** that combines hardware devices, APIs, and online services in a revolutionary way on a browser. It provides the following features:

- Browser-based UI.

- Works with Node.js and is lightweight.

- Encapsulates function and can be used as a node (meaning functions are locked in an abstract capsule) .

- You can create and add your own nodes.

- Easy access to IBM Cloud services.

In other words, it can be said that this tool is suitable for building IoT-related services, such as data control on devices, and linking edge devices and cloud services. Originally, the development concept of Node-RED was for IoT, so this makes sense.

Now, let's look at the basic structure of IoT so that those who are only vaguely aware of IoT can understand it. It can be said that IoT is basically composed of six layers, as shown in the following diagram:

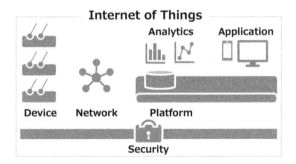

Figure 1.6 – IoT six layers

Let's take a look at these in more detail.

Device

The device is a so-called edge device. IoT has various sensors and handles the data that's acquired from them. Since it doesn't make sense to have the data only on the edge device, we need to send that data through the gateway to the network.

Network

This is the network that's required to send the data that's been obtained from the device to a server on the internet. It usually refers to the internet. In addition to the internet, there is also a P2P connection via Bluetooth or serial.

Platform

The party that receives and uses the data is the platform. We may also have a database for activating and authenticating things, managing communications, and persisting received data.

Analytics

This is a layer that analyzes the received data. Broadly speaking, it may be classified as an application. This is the part that prepares the data so that it can be processed into a meaningful form.

Application

An application provides a specific service based on data analysis results. It can be a web or mobile application, or it can be a hardware-specific embedded application. It can be said to be the layer that's used by the end user of the IoT solution.

Now that we have an understanding of IoT, we will examine why Node-RED should be used for it.

Node-RED and IoT

While explaining IoT so far, we've made it clear why Node-RED is suitable for IoT. For example, you can understand why FBP tools that have been developed for IoT survive when used with Node-RED. In particular, the following three points should be taken into account:

- Since it can be run on edge devices (pre-installed on specific versions of Raspberry Pi OS), it is ideal for data handling at the device layer.

- Since it can be run on the cloud (provided as a default service in IBM Cloud), it is easy to link with storage and analysis middleware.

- Since MQTT and HTTP protocols can be covered, it is very easy to exchange data between the edge device and the server processing cloud.

In this way, Node-RED, which largely covers the elements required for IoT, is now used for a wide range of applications, such as web services and chart display, as well as programming for IoT. Also, as of June 2020, if you look at Google Trends for Node-RED, you can see that the number of users is gradually increasing. As such, Node-RED is a very attractive FBP tool:

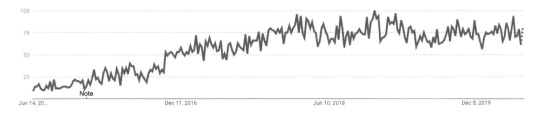

Figure 1.7 – Google Trends for "Node-RED"

A typical edge device that can use Node-RED is Raspberry Pi. Of course, it is possible to use Node-RED on other platforms, but it goes well with Raspberry Pi, which also has a pre-installed version of the OS.

> **Raspberry Pi OS Supports Node-RED**
>
> Node-RED has also been packaged for the Raspberry Pi OS repositories and appears in their list of recommended software. This allows it to be installed using `apt-get install Node-RED` and includes the Raspberry Pi OS-packaged version of Node.js, but does not include npm. More information can be found at `https://nodered.org/docs/getting-started/raspberrypi`.

IBM Cloud is a typical cloud platform that can use Node-RED. Of course, you can use Node-RED on other clouds, but IBM Cloud provides a service that anyone can easily start.

> **Important Note**
>
> Node-RED is available on the IBM Cloud platform as one of its Starter Kits applications in their catalog. It is very easy to start using the flow editor as a web application on IBM Cloud (`https://nodered.org/docs/getting-started/ibmcloud`).

Summary

In this chapter, you learned what FBP and Node-RED are. Due to this, you now understand why Node-RED is currently loved and used by lots of people as an FBP tool. At this point, you may want to build an application using Node-RED. In the next chapter, we'll install Node-RED in our environment and take a look at it in more depth.

2
Setting Up the Development Environment

In this chapter, you will install the tools that you'll need to use Node-RED. This extends not only to Node-RED itself, but to its runtime, Node.js, and how to update both Node-RED and Node.js.

Node-RED released its 1.0 milestone version in September 2019. This reflects the maturity of the project, as it is already being widely used in production environments. It continues to be developed and keeps up to date by making changes to the underlying Node.js runtime. You can check the latest status of Node-RED's installation at `https://nodered.org/docs/getting-started/`.

There are a number of installation guides on the Node-RED official website, such as local, Raspberry Pi, Docker, and major cloud platforms.

In this chapter, you will learn how to install Node-RED on your local computer, whether you are running it on Windows, Mac, or on a Raspberry Pi. We will cover the following topics:

- Installing npm and Node.js for Windows
- Installing npm and Node.js for Mac
- Installing npm and Node.js for Raspberry Pi
- Installing Node-RED for Windows
- Installing Node-RED for Mac
- Installing Node-RED for Raspberry Pi

By the end of this chapter, we'll have all the necessary tools installed and be ready to move on to building some basic flows with Node-RED.

For reference, the author's test operation environment is Windows 10 2004 18363.476, macOS Mojave 10.14.6 (18G5033), and Raspberry Pi OS 9.4 stretch.

Technical requirements

You will need to install the following for this chapter:

- Node.js (v12.18.1)*
- npm (v6.14.5)*

*LTS version at the time of writing for both.

Installing npm and Node.js for Windows

If you want to use Node-RED on Windows, you must install npm and Node.js via the following website:

`https://nodejs.org/en/#home-downloadhead`.

You can get the Windows Installer of Node.js directly there. After that, follow these steps:

1. Access the original Node.js website and download the installer.

    You can choose both versions – **Recommended** or **Latest Features** – but in this book, you should use the **Recommended** version:

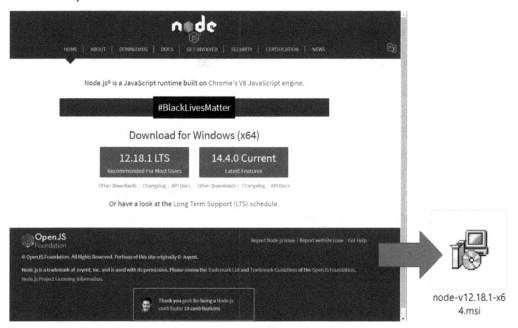

Figure 2.1 – Choosing a Recommended version installer

2. Click the msi file you downloaded to start installing Node.js. It includes the current version of npm. Node-RED is running on the Node.js runtime, so it is needed.

3. Simply click the buttons of the dialog windows according to the installation wizard, though there are some points to bear in mind during the install.

4.   Next, you need to accept the End-User License Agreement:

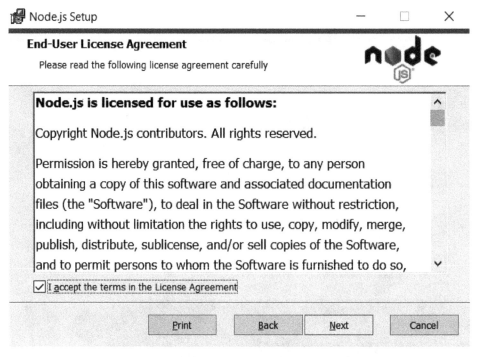

Figure 2.2 – End-User License Agreement window

You can also change the install destination folder. In this book, the default folder
(C:/Program Files/nodejs/) will be used:

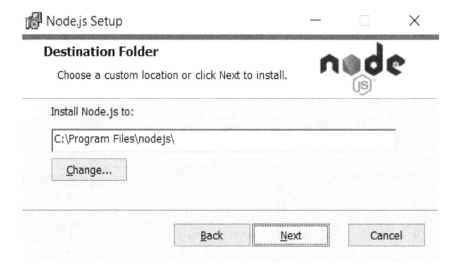

Figure 2.3 – Installing the destination folder

5. No custom setup is needed on the next screen. You can select **Next** with only the default features selected:

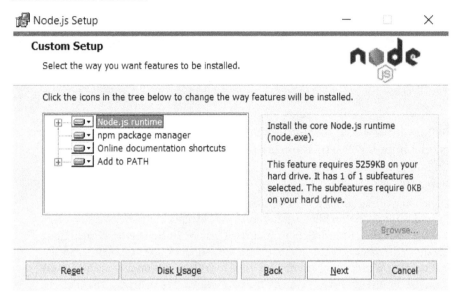

Figure 2.4 – No custom setup is needed

6. On the following screen, you can click **Next** without checking anything. However, it's OK to install the tools that can be selected here. This includes the installations and settings the path of these environments (Visual C++, windows-build-tools, and Python):

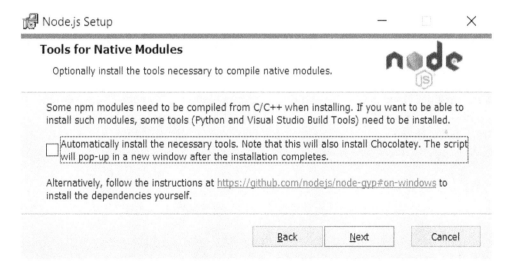

Figure 2.5 – Tools for Native Modules window

7. Check the versions of your tools with the following commands when the installation for Node.js has finished:

```
$ node --version
v12.18.1


$ npm -version
6.14.5
```

When the installations of Node.js and npm are complete, you can check their version numbers. With this, you are prepared to install Node-RED.

> **Important note**
>
> Depending on the project, it is common for the operation to be stable with the old Node.js version but for it not to work if you use a different version of Node.js. However, uninstalling your current version of Node.js and installing the desired version of Node.js every time you switch projects takes time. So, if you're using Windows, I recommend using a Node.js version management tool such as nodist (https://github.com/nullivex/nodist). There are other kinds of version control tools for Node.js, so please try to find one that is easy for you to use.

Installing npm and Node.js for Mac

If you want to use Node-RED on macOS, you must install npm and Node.js via the following website:

https://nodejs.org/en/#home-downloadhead

You can get the Mac Installer for Node.js directly there.

Access the original Node.js website and download the installer. You can choose either the recommended or latest features version, but for this book, you should use the recommended version:

Figure 2.6 – Choosing a recommended version installer

Click the .pkg file you downloaded to start installing Node.js. It includes the current version of npm. Node-RED is running on the Node.js runtime, so it is needed. Simply click according to the installation wizard, though there are some points in the installation to pay attention to.

You need to accept the End-User License Agreement:

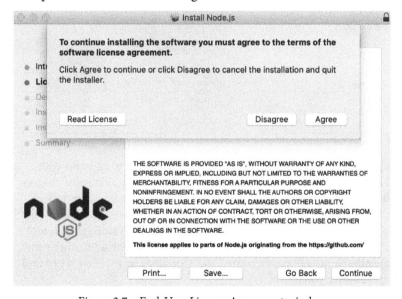

Figure 2.7 – End-User License Agreement window

You can change the installation location. In this book, the default location (Macintosh HD) will be used:

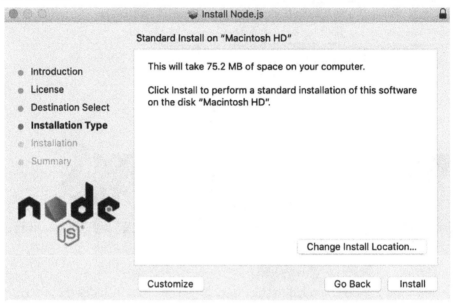

Figure 2.8 – Install location

You can check the versions of your tools with the following commands when the installation for Node.js has finished. Once you have finished installing Node.js and npm, you can check their version numbers. You have already prepared to install Node-RED:

```
$ node --version
v12.18.1


$ npm -version
6.14.5
```

> **Note**
>
> Depending on the project, it is common for the operation to be stable with the old Node.js version, and that it will not work if you use a different version of Node.js. However, uninstalling the current Node.js version and installing the desired version of Node.js every time you switch projects takes time. So, if you're using macOS, I recommend using a Node.js version management tool such as Nodebrew (https://github.com/hokaccha/nodebrew). There are other kinds of version control tools for Node.js, so please try to find one that is easy for you to use.

Now that we have covered the installation processes for both Windows and Mac, let's learn how to install npm and Node.js for Raspberry Pi.

Installing npm and Node.js for Raspberry Pi

If you want to use Node-RED on Raspberry Pi, congratulations – you are already prepared to install Node-RED. This is because Node.js and npm are installed by default. You can use the existing installation script to install Node-RED, including Node.js and npm. This script will be described later in this chapter, in the *Installing Node-RED for Raspberry Pi* section, so you can skip this operation for now.

However, you should check your Node.js and npm versions on your Raspberry Pi. Please type in the following commands:

```
$ node --version
v12.18.1


$ npm –version
6.14.5
```

If it is not the LTS version or stable version, you can update it via the CLI. Please type in and run the following commands to do this. In this command, on the last line, lts has been used, but you can also put stable instead of lts if you want to install the stable version:

```
$ sudo apt-get update
$ sudo apt-get install -y nodejs npm
$ sudo npm install npm n -g
$ sudo n lts
```

Now that we have successfully checked the versions of Node.js and npm on our Raspberry Pi and updated them (if applicable), we will move on to installing Node-RED for Windows.

> **Important note**
> The script the Node-RED project provides takes care of installing Node.js and npm. It is not generally recommended to use the versions that are provided by Raspberry Pi OS due to the strange ways they package them.

Installing Node-RED for Windows

In this section, we will explain how to set up Node-RED in a Windows environment. This procedure is for Windows 10, but it will work for Windows 7 and Windows Server 2008 R2 and above as well. Windows 7 or earlier versions of Windows Server 2008 R2 are not currently supported and are not recommended.

For Windows, installing Node-RED as a global module adds the node-red command to your system path. Run the following command in Command Prompt:

```
$ npm install -g --unsafe-perm node-red
```

Once you have finished installing Node-RED, you can use Node-RED straight away. Please run the following command. After running this command, you will recognize the URL being used to access the Node-RED flow editor. Usually, localhost (127.0.0.1) with the default port 1880 will be allocated:

```
$ node-red
Welcome to Node-RED
===================
...
[info] Starting flows
[info] Started flows
[info] Server now running at http://127.0.0.1:1880/
```

Let's access Node-RED on a browser. For this, type in the URL you received from Command Prompt. I strongly recommend using Chrome or Firefox for running Node-RED:

Figure 2.9 – Node-RED flow editor

Now, you are ready to program in Node-RED. From *Chapter 3*, *Understanding Node-RED Characteristics by Creating Basic Flows*, onward, we will learn how to actually build an application using Node-RED.

For now, let's move on to installing Node-RED in macOS.

Installing Node-RED for Mac

In this section, we will explain how to set up Node-RED in a macOS environment. This procedure is for macOS Mojave. It will likely work for all versions of Mac OS X, but I strongly recommend that you use the current version of macOS.

For macOS, installing Node-RED as a global module adds the node-red command to your system path. Run the following command in the Terminal. You may need to add sudo at the front of the command, depending on your local settings:

```
$ sudo npm install -g --unsafe-perm node-red
```

You can also install Node-RED with other tools. This is mainly for Mac/Linux or the kinds of OS that support the following tools:

1. Docker (`https://www.docker.com/`), if you have the environment for running Docker.

    The current Node-RED 1.x repository on Docker Hub has been renamed "nodered/node-red".

    Versions up to 0.20.x are available from `https://hub.docker.com/r/nodered/node-red-docker`.

    > **Important note**
    > When running Node-RED with Docker, you need to ensure that the added nodes and flows will not be lost if the container breaks. This user data can be persisted by mounting the data directory to a volume outside the container. You can also do this by using a bound mount or a named data volume.

    Run the following command to install Node-RED with Docker:

    ```
    $ docker run -it -p 1880:1880 --name mynodered nodered/
    node-red
    ```

2. Snap (`https://snapcraft.io/docs/installing-snapd`) if your OS supports it.

    If you install it as a Snap package, you can run it in a secure container that doesn't have access to the external features you have to use, such as the following:

    * Access main system storage (only read/write to local home directory is allowed).
    * Gcc: Required to compile the binary components for the node you want to install.
    * Git: Required if you want to take advantage of project features.
    * Direct access to GPIO hardware.
    * Access to external commands, such as flows executed in Exec nodes.

    There's less security for containers, but you can also run them in **classic** mode, which gives you more access.

    Run the following command to install Node-RED with Snap:

```
$ sudo snap install node-red
```

Once you have finished installing Node-RED, you can use Node-RED immediately. Please run the following command. After running this command, you can find the URL for accessing the Node-RED flow editor. Usually, localhost (127.0.0.1) with the default port 1880 will be allocated:

```
$ node-red
Welcome to Node-RED
===================
...
[info]  Server now running at http://127.0.0.1:1880/
[info]  Starting flows
[info]  Started flows
```

Let's access Node-RED on a browser. Type in the URL you received from Command Prompt. I strongly recommend using Chrome or Firefox for running Node-RED:

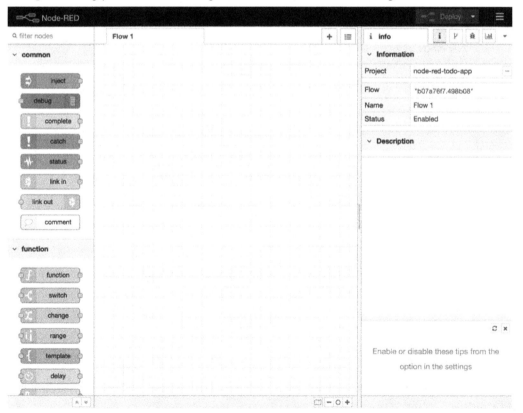

Figure 2.10 – Node-RED flow editor

Now, you are ready to program in Node-RED. In *Chapter 3*, *Understanding Node-RED Characteristics by Creating Basic Flows,* we will learn how to actually build an application using Node-RED.

Our final installation will be for Node-RED on Raspberry Pi.

Installing Node-RED for Raspberry Pi

In this section, we will explain how to set up Node-RED in a Raspberry environment. This procedure is for Raspberry Pi OS Buster (Debian 10.x), but it will work for Raspberry Pi OS Jessie (Debian 8.x) and above.

You can check your version of Raspberry Pi OS easily. Just run the following command on your Terminal:

```
$ lsb_release -a
```

If you want to also check the version of Debian you have, please run the following command:

```
$ cat /etc/debian_version
```

You have now prepared to install Node-RED. The following script installs Node-RED, including Node.js and npm. This script can also be used for upgrading your application, which you have already installed.

> **Note**
> This instruction is subject to change, so it is recommended that you refer to the official documentation as needed.

This script works on Debian-based operating systems, including Ubuntu and Diet-Pi:

```
$ bash <(curl -sL https://raw.githubusercontent.com/node-red/
linux-installers/master/deb/update-nodejs-and-nodered)
```

You may need to run sudo apt install build-essential git to ensure that npm can build the binary components that need to be installed.

Node-RED is already packaged as a Raspberry Pi OS repository and is included in the *Recommended Software* list. It can be installed with the apt-get install Node-RED command, and it also contains a Raspberry Pi OS packaged version of Node.js, but npm is not included.

While using these packages may seem convenient at first glance, it is highly recommended to use the installation script instead.

After the installation, you can start Node-RED and access the Node-RED flow editor. We have two ways to start it, as follows:

1. Run with the CLI: If you want to run Node-RED locally, you can start Node-RED by using the `node-red` command in your Terminal. Then, you can stop it by pressing *Ctrl* + *C* or closing the Terminal window:

    ```
    $ node-red
    ```

2. Run via Programming menu: Once Node-RED has been installed, you can start it from the Raspberry Pi menu. Click **Menu | Programming | Node-RED** to open the Terminal and launch Node-RED. Once Node-RED has been launched, you can access the Node-RED flow editor from your browser, just as you would in the CLI:

Figure 2.11 – Accessing Node-RED via the Raspberry Pi menu

After launching Node-RED from the menu, you should check the Node-RED running process on your Terminal and find the URL of the Node-RED flow editor. It is usually the same URL as the one that can be launched via the CLI directly:

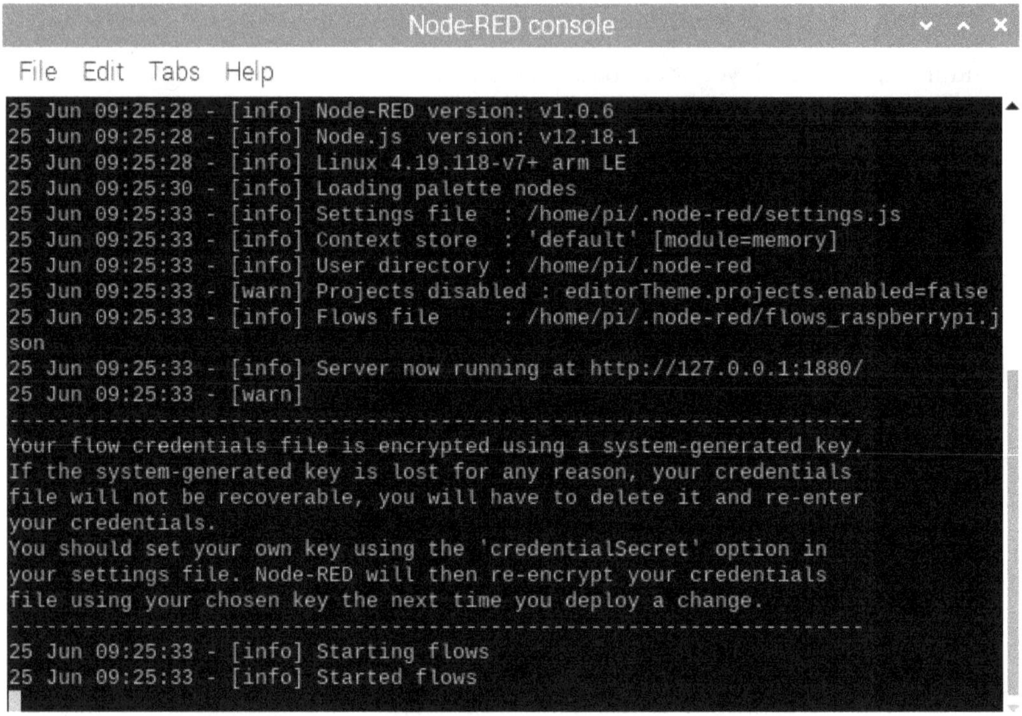

```
Node-RED console                                              v  ^  x

File  Edit  Tabs  Help

25 Jun 09:25:28 - [info] Node-RED version: v1.0.6
25 Jun 09:25:28 - [info] Node.js  version: v12.18.1
25 Jun 09:25:28 - [info] Linux 4.19.118-v7+ arm LE
25 Jun 09:25:30 - [info] Loading palette nodes
25 Jun 09:25:33 - [info] Settings file  : /home/pi/.node-red/settings.js
25 Jun 09:25:33 - [info] Context store  : 'default' [module=memory]
25 Jun 09:25:33 - [info] User directory : /home/pi/.node-red
25 Jun 09:25:33 - [warn] Projects disabled : editorTheme.projects.enabled=false
25 Jun 09:25:33 - [info] Flows file     : /home/pi/.node-red/flows_raspberrypi.j
son
25 Jun 09:25:33 - [info] Server now running at http://127.0.0.1:1880/
25 Jun 09:25:33 - [warn]

---------------------------------------------------------------------
Your flow credentials file is encrypted using a system-generated key.
If the system-generated key is lost for any reason, your credentials
file will not be recoverable, you will have to delete it and re-enter
your credentials.
You should set your own key using the 'credentialSecret' option in
your settings file. Node-RED will then re-encrypt your credentials
file using your chosen key the next time you deploy a change.
---------------------------------------------------------------------

25 Jun 09:25:33 - [info] Starting flows
25 Jun 09:25:33 - [info] Started flows
```

Figure 2.12 – Checking the URL to access the Node-RED flow editor

Let's access Node-RED on a browser. You can type in the URL you received from the Command Prompt to do this. If your Raspberry Pi default web browser is Chromium, then there should be no problems with using Node-RED. However, if you wish to use another browser, I strongly recommend installing Chromium for running Node-RED:

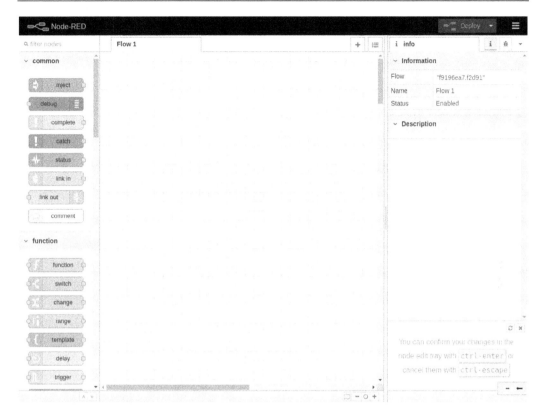

Figure 2.13 – Node-RED flow editor

And that's it! We have now covered all the installation options for each tool we'll need in order to start using Node-RED.

Summary

In this chapter, you've gotten your environment ready so that you can use the Node-RED flow editor. At this point, I believe that you can already access the Node-RED flow editor, so you'll want to learn how to use it. In the next chapter, we'll make a sample flow on it and learn about the major features of the Node-RED flow editor.

3
Understanding Node-RED Characteristics by Creating Basic Flows

In this chapter, we'll actually create a flow using Node-RED Flow Editor. By creating a simple flow, you will understand how to use the tool and its characteristics. For a better understanding, we will create some sample flows.

From now on, you will create applications called flows using Node-RED. In this chapter, you will learn how to use Node-RED and how to create an application as a flow. To do this, we will cover the following topics:

- Node-RED Flow Editor mechanisms
- Using the Flow Editor
- Making a flow for a data handling application

- Making a flow for a web application

- Importing and exporting a flow definition

By the end of this chapter, you will have mastered how to use Node-RED Flow Editor and know how to build a simple application with it.

Technical requirements

To complete this chapter, you will need the following:

- Node-RED (v1.1.0 or above).

- The code for this chapter can be found in Chapter03 folder at https://github.com/PacktPublishing/-Practical-Node-RED-Programming.

Node-RED Flow Editor mechanisms

As you learned in the previous chapters, Node-RED has two logical parts: a development environment called the Flow Editor and an execution environment for executing the application that's been created there. These are called the runtime and the editor, respectively. Let's take a look at them in more detail:

- **Runtime**: This includes a Node.js application runtime. It is responsible for running the deployed flows.

- **Editor**: This is a web application where the user can edit their flows.

The main installable package includes both components, with a web server to provide Flow Editor as well as a REST Admin API for administering the runtime. Internally, these components can be installed separately and embedded into existing Node.js applications, as shown in the following diagram:

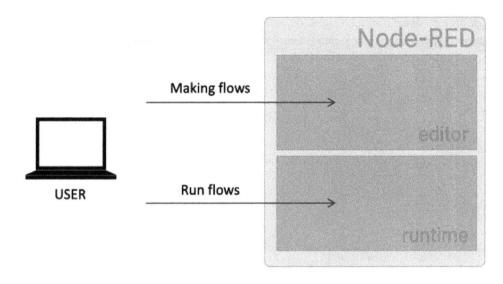

Figure 3.1 – Node-RED overview

Now that you understand the mechanisms of Node-RED, let's immediately learn how to use the Flow Editor.

Using the Flow Editor

Let's take a look at the main functions of the Flow Editor.

The main features of the Flow Editor are as follows:

- **Node**: The main building block of Node-RED applications, they represent well-defined pieces of functionality.

- **Flow**: A series of nodes wired together that represent the series of steps messages pass through within an application.

- **The panel on the left is the palette**: A collection of nodes that are available within the editor that you can use to build your application.

- **Deploy button**: Press this button to deploy your apps once you've edited them.

- **Sidebar**: A panel for displaying various functions, such as processing parameter settings, specifications, and debugger display.

- **Sidebar tabs**: Settings for each node, standard output, change management, and so on.

- **Main menu**: Flow deletion, definition import/export, project management, and so on.

These functions are arranged on the screen of the Flow Editor like so:

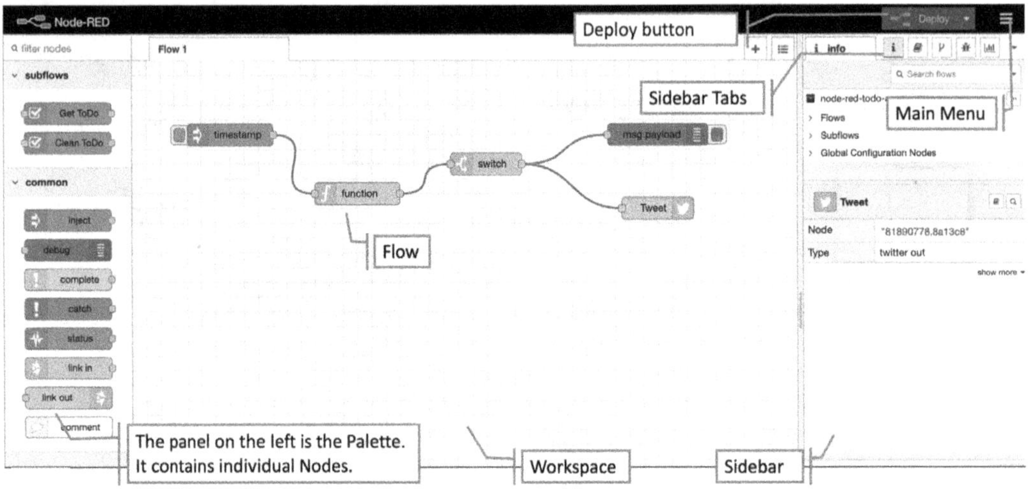

Figure 3.2 – Node-RED Flow Editor

You need to understand what is contained in the Flow menu before you start using Node-RED. Its contents may differ, depending on the version of Node-RED you're using, but it has some setting items such as **Project management of flow**, **Arrange view**, **Import / export of flow**, **Installation of node published in library**, and so on that are universal. For more information on how to use Node-RED, it's a good idea to refer to the official documentation as needed.

> **Important note**
> Node-RED User Guide: https://nodered.org/docs/user-guide/.

The following diagram shows all these Flow Editor menu options inside Node-RED:

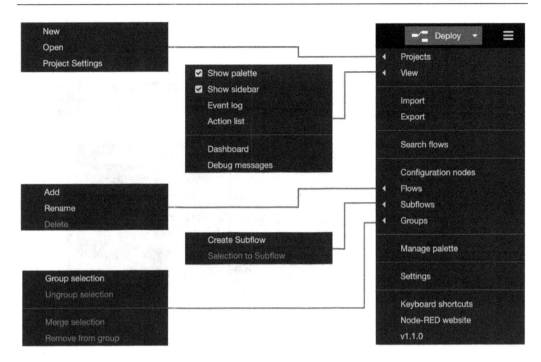

Figure 3.3 – Node-RED Flow Editor menu

With that, you are ready to use Node-RED to build an application. So, let's get started!

First of all, you need to run Node-RED in your environment. Please refer to *Chapter 2, Setting Up the Development Environment*, to learn how to set it up with your environment, such as Windows, Mac, or Raspberry Pi, if you haven't done so already.

With Node-RED running, let's move on to the next section, where we'll be making our first flow.

Making a flow for a data handling application

In this section, you will create a working application (called a flow in Node-RED). Whether it is the **internet of things (IoT)** or server processing as a web application, the basic operation that Node-RED performs is sequentially transferring data.

Here, we'll create a flow where JSON data is generated in a pseudo manner, and the data is finally output to standard output via some nodes on Node-RED.

There are many nodes on the left-hand side of the palette. Please pay attention to the **common** categories here. You should be able to easily find the **inject** node, as shown in the following screenshot:

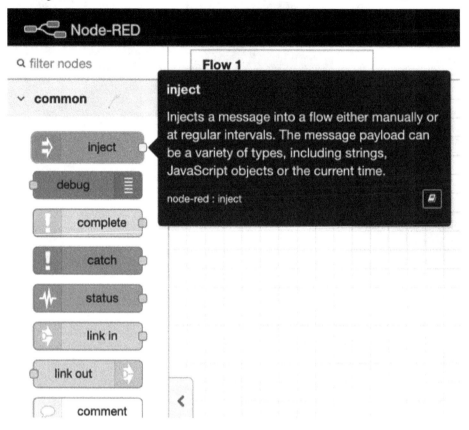

Figure 3.4 – Inject node

This node can inject a message into the next node. Let's get started:

1. Drag and drop it onto the palette of Flow 1 (the default flow tab).

    You will see that the node is labeled with the word **timestamp**. This is because its default message payload is a timestamp value. We can change the data type, so let's change it to a JSON type.

2. Double-click the node and change its settings when the **Properties** panel of the node is opened:

Figure 3.5 – Edit inject node Properties panel

3. Click the drop-down menu of the first parameter and select {}**JSON**. You can edit the JSON data by clicking the […] button on the right-hand side.

4. Click the […] button, and the JSON editor will open. You can make JSON data with a text-based editor or a visual editor.

5. This time, let's make JSON data with an item called {"name" : "Taiji"}. You should replace my name with your name:

Figure 3.6 – JSON editor

Great – you have successfully made some sample JSON data!

6. Click the **Done** button and close this panel.

7. Similarly, place a **Debug** node on the palette.

8. After placing it, wire the **Inject** and **Debug** nodes to it.

    Once you execute this flow, the JSON data that was passed from the **Inject** node will be output to the debug console (standard output) by the **Debug** node. You don't need to configure anything on the **Debug** node:

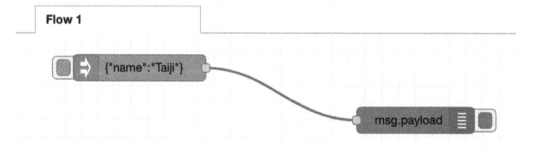

Figure 3.7 – Placing the Debug node and wiring it

9. Finally, you need to deploy the flow you created. In Node-RED Flow Editor, we can deploy all our flows on the workspace to the Node-RED runtime by clicking the **Deploy** button in the top-right corner.

10. Before running the flow, you should select the **Debug** tab from the node menu's side panel to enable the debug console, as shown in the following screenshot:

Figure 3.8 – Enabling the debug console

11. Let's run this flow. Click the switch of the **Inject** node to see the result of executing the flow on the debug console:

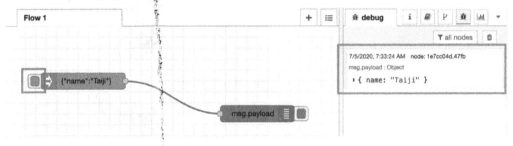

Figure 3.9 – Executing the flow and checking the result

This is a very simple and easy data handling flow sample. In the latter half of this book, we will also experiment with data handling by actually connecting IoT devices and passing data obtained from a web API. In this section, it is enough that you understand how to handle data in Node-RED. Next, we're going to experiment with making a flow for a web application.

Making a flow for a web application

In this section, you will create a new flow for a web application. We'll create this flow in the same way we created the previous data handling flow.

You can create it in the workspace of the same flow (Flow 1), but to make things clear and simple, let's create a new workspace for the flow by following these steps:

1. Select **Flows | Add** from the **Flow** menu. Flow 2 will be added to the right-hand side of Flow 1. These flow names, such as "Flow 1" and "Flow 2," are default names that are provided upon creation. You can rename the flow so that it has a more specific name if you want to:

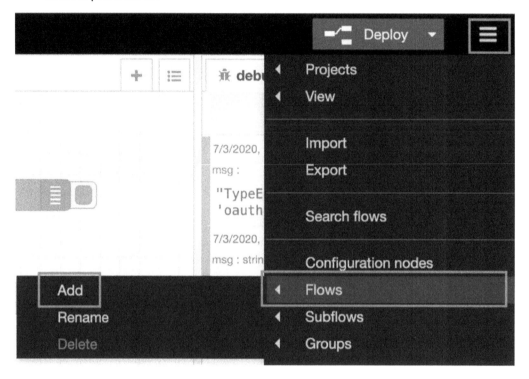

Figure 3.10 – Adding a new flow

2. Select the **http in** node from the **network** category on the palette, and then drag and drop it onto the palette of Flow 2 (the new flow tab you just added):

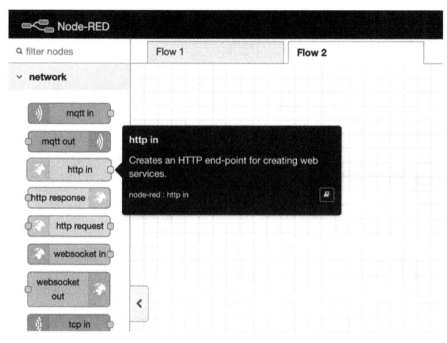

Figure 3.11 – An http in node

3. Double-click the node to open its **Edit** dialog.

4. Enter the URL (path) of the web application you will create.

   This path will be used as part of the URL for the web application you will be creating, under the Node-RED URL. In this case, if your Node-RED URL is `http://localhost:1880/`, your web application URL will be `http://localhost:1880/web`. An example of this can be seen in the following screenshot:

Figure 3.12 – Setting the path of the URL

5. To send a request via HTTP, an HTTP response is required. So, place an **http response** node on the workspace of your Node-RED.

    You can find this node in the **network** category of the palette, next to the **http in** node. Here, the **http response** node simply returns the response, so you don't need to open the configuration panel. You can leave it as-is. If you want to include a status code in the response message, you can do so from the **settings** panel, as shown in the following screenshot:

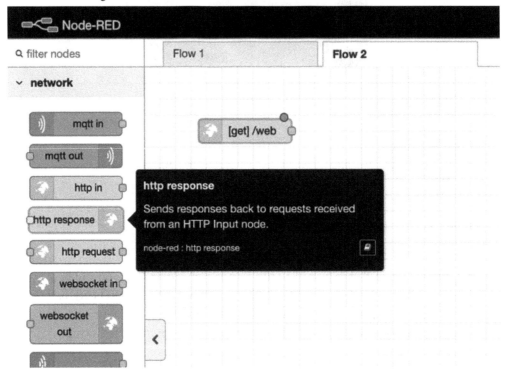

Figure 3.13 – An http response node

6. After placing an **http response** node on the palette, add a wire from the **http in** node to the **http response** node.

This completes the flow for the web application, since we've allowed an HTTP request and response. You will see a light blue dot in the top-right corner of each node, which indicates that they haven't been deployed yet – so please make sure you click the **Deploy** button:

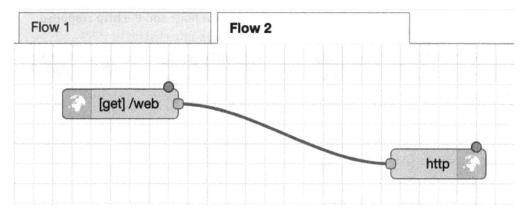

Figure 3.14 – Wired nodes

7. Once it's been successfully deployed, open a new tab in your browser.

8. Then, access the URL of the web application shown in the **http in** node section by entering `http://localhost:1880/web`.

You should find that only {} is displayed on your screen. This is not a mistake. It is a result of sending an HTTP request and returning a response to it. Right now, since we have not set the content to be passed to the response, an empty JSON is passed as message data. This looks as follows:

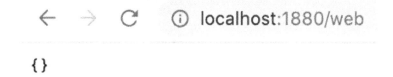

{}

Figure 3.15 – Web application result

This isn't great, so let's create some content. Let's do something very simple and implement some simple HTML code. So, where should I code this? The answer is simple. Node-RED has a template node that allows you to specify the HTML code as-is as output. Let's use this:

1. Drag and drop a **template** node between the **http in** node and the **http response** node on the wire, so that the **template** node will be connected on it:

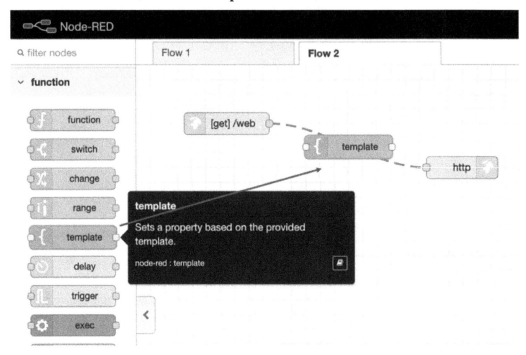

Figure 3.16 – Placing a "template" node on the wire between our two existing nodes

2. Next, double-click the **template** node to open the settings panel. You can code on the **Template** area of the **settings** panel. This time, use the following sample HTML. The title is specified for the head. Let's attach a page heading to the body with the <h1> tag. Arrange the contents resembling the menu with the <h2> tag. The code will look like this:

```
<html>
  <head>
    <title>Node-RED Web sample</title>
  </head>
  <body>
    <h1>Hello Node-RED!!</h1>
```

```
    <h2>Menu 1</h2>
    <p>It is Node-RED sample webpage.</p>
    <hr>
    <h2>Menu 2</h2>
    <p>It is Node-RED sample webpage.</p>
  </body>
</html>
```

> **Note**
>
> You can also get this code from this book's GitHub repository at `https://github.com/PacktPublishing/-Practical-Node-RED-Programming/tree/master/Chapter03`.

3. Once you have finished editing the **template** node, click the **Done** button to close it.

    The following screenshot shows what your template node will look like as you edit it:

Figure 3.17 – Code in the Template area

With that, we have finished preparing the HTML to be shown on our page. Please make sure you click the **Deploy** button. Access the web page by going to `http://localhost:1880/web` once more. You should now see the following output:

Hello Node-RED!!

Menu 1

It is Node-RED sample webpage.

Menu 2

It is Node-RED sample webpage.

Figure 3.18 – Web application result

At this point, you should understand how to make a web application on Node-RED. I imagine it has been nice and easy so far. Now that we have built up some momentum, let's continue learning. In the next section, we will import and export the flow definition that we have created.

Importing and exporting a flow definition

In this section, you will import and export the flow definition you have created. Usually, when developing, it is necessary to back up the source code and version control. You may also import source code created by others, or export your own source code and pass it on to others. Node-RED has a similar concept. In Node-RED, it is a normal practice to import and export the flow itself instead of importing or exporting the source code (for example, the template node described previously).

So, first, let's export the flow we have created so far. This is easy to do:

1. Simply select **Export** from the **Edit** dialog under the **Main** menu of the Node-RED Flow Editor.

    When the **Export** menu is displayed, you can only select the current flow or all your flows. You can also select raw JSON, without indentation, or formatted JSON, with indentation.

2. Here, select the current flow and select **Formatted**.

3. Now, you can select how to save the exported JSON data – **Copy to clipboard** or **Download**. Here, we'd want to download the JSON data, so click the **Download** button:

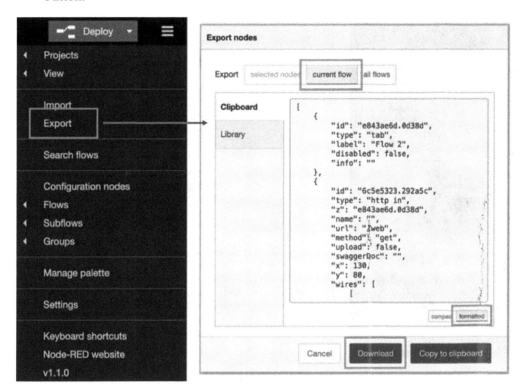

Figure 3.19 – Export operation

You will see a file called flows.json in the downloads location of your machine.

4. Open this file in a text editor so that you can check the contents of the JSON file.

With that, we have learned how to export.

Next, we need to import this definition (flows.json) into our Node-RED Flow Editor. Do this by following these steps:

1. Simply select **Import** from the **Flow** menu in the Node-RED Flow Editor.

    When the **Import** menu is displayed, you can select **Paste flow json** or **Select a file based import**. You can also select a **current flow** or a **new flow** from the flow tab. If you select **new flow**, a new flow tab will be added automatically.

2. Here, please choose **Select a file based import** and import to **new flow**. Then, pick the JSON file called flows.json you exported to your local machine.

3.   Once the file has loaded, click the **Import** button:

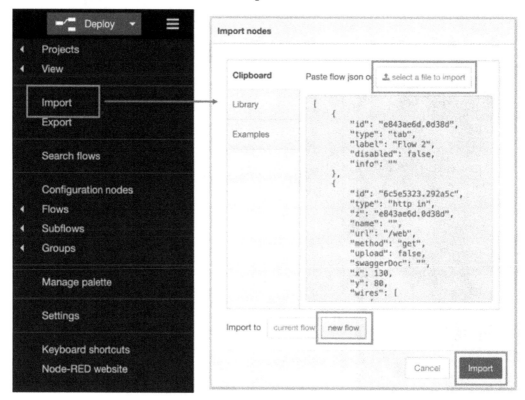

Figure 3.20 – Import operation

4.   You now have the new tab, named Flow 2, next to the same flow on the old Flow 2 tab. It has been imported completely, but it hasn't been deployed yet, so click the **Deploy** button, as follows:

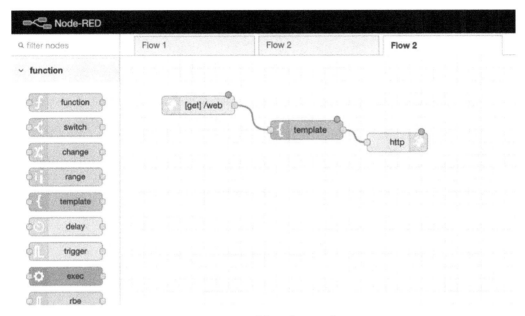

Figure 3.21 – Adding the new flow

With that, we've successfully prepared what will be shown on our web page using the flow we imported. Please make sure you click **Deploy** button.

5.   Access the web page again by going to http://localhost:1880/web.

Here, you will see that this web page has the same design as the web page you exported. Great work!

Figure 3.22 – Result of the web application

Now, let's wrap this chapter up.

Summary

In this chapter, you learned how to use Node-RED Flow Editor to make basic flows and import/export flows. Now that you know how to use Node-RED Flow Editor, you'll want to learn about more of its features. Of course, Node-RED doesn't only have basic nodes such as **Inject**, **http**, and **template**, but also more attractive nodes such as **switch**, **change**, **mqtt**, and **dashboard**. In the next chapter, we'll try to use several major nodes so that we can code JavaScript, catch errors, perform data switching, delay functions, use the CSV parser, and more.

4
Learning the Major Nodes

In this chapter, you will learn about the major nodes used in Node-RED. Node-RED, which is an open source project, provides some major nodes by default, but it is possible to import and use nodes from the public library as required.

Node-RED has a lot of nodes. Therefore, this book is not sufficient to explain all of them. So, in this chapter, let's pick up the main nodes and most commonly used basic nodes and learn how to use them, exploring these topics in this chapter:

- What is a node?
- How to use nodes
- Getting various nodes from the library

By the end of this chapter, you will have mastered how to use major nodes in the Node-RED flow editor.

Technical requirements

To progress in this chapter, you will need the following technical requirements:

- Node-RED (v1.1.0 or above).
- The code used in this chapter can be found in `Chapter04` folder at `https://github.com/PacktPublishing/-Practical-Node-RED-Programming`.

What is a node?

Let's first understand what exactly a node is in Node-RED.

Node-RED is a tool for programming Node.js applications with **Graphical User Interface (GUI)** tools. Node-RED also serves as an environment for executing software (Node-RED Flow) programmed on Node-RED.

Normally, when programming with Node.js, the source code is written with a code editor or **Integrated Development Environment (IDE)**. An executable file is generated by building the written source code (compiling, associating with dependency files, and so on).

Visual programming on Node-RED basically follows the same process. The difference is that the coding part is the act of placing the node on Node-RED instead of the editor.

In Node-RED, the basic processing used when programming with Node.js is provided by implemented parts called nodes. In normal object-oriented programming, these parts may often be provided as library files in the form of common parts.

Since Node-RED is a GUI-based visual programming tool, these common parts are more than just library files. These common parts are shaped like boxes and are called nodes in Node-RED. Also, except for some nodes, generally nodes can set the things that can be variables (arguments, parameters, and so on) as node properties when programming.

In other words, since there are already programmed parts (nodes), programming is completed simply by placing them in the GUI. The following figure compares pure Node.js programming with flow creation in Node-RED:

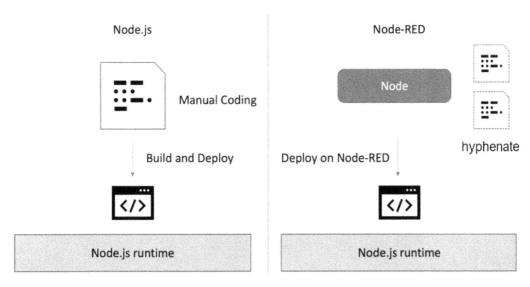

Figure 4.1 – Node-RED versus Node.js programming

Now that you understand the concepts of Node-RED and nodes, let's take a closer look at nodes.

As you can see when you start Node-RED, the basic processing nodes are provided in the Node-RED flow editor by default. This is called a **pre-installed node**.

The following are typical categories of pre-installed nodes:

- **Common**: This includes nodes that inject specific data into the flow, nodes that judge the processing status, and nodes that output logs for debugging.

- **Function**: This includes nodes that can write directly in JavaScript and HTML, nodes that convert parameter variables, and nodes that make conditional branches depending on the contents of those parameters.

- **Network**: This includes nodes that handle the protocol processing required for communication, such as MQTT, HTTP, and WebSockets.

Of course, the examples given here are just a few. There are actually many more categories and nodes.

> **Important note**
> The pre-installed nodes also depend on the Node-RED version. It's a good idea to check the official documentation for information on your Node-RED version: `https://nodered.org/docs/`.

Nodes are arranged like parts on the Node-RED flow editor and can be used simply by connecting up the wiring. As mentioned earlier, you don't have to code it yourself, except for some nodes.

Basically, the flow editor has the appearance of a box and has a settings window inside it. In the settings window, you can set the required parameters and configurations for each node:

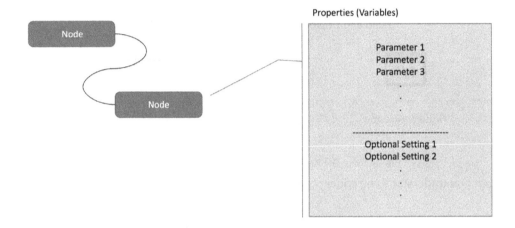

Figure 4.2 – Nodes

That's all the concepts you need to know about nodes. In the next section, you will learn how to actually use nodes.

How to use nodes

In this section, we will learn how to use nodes.

Visual programming in Node-RED is a little different from other visual programming tools because it uses flow-based programming. But rest assured, it's not difficult at all. If you actually create a few simple flows, you should be able to master how to use nodes in Node-RED.

So, let's now create a sample flow using some typical preinstalled nodes. The environment is the same for Raspberry Pi, Windows, and macOS systems. Please use your favorite environment.

Common category

Let's introduce the nodes that we'll use to make our flow. You can pick all of the nodes up and place them on the palette from the common category.

Create a sample flow with nodes in the common category. The following four nodes are used:

- The **inject** node
- The **complete** node
- The **catch** node
- The **debug** node

Place and wire up the nodes as shown in the following figure:

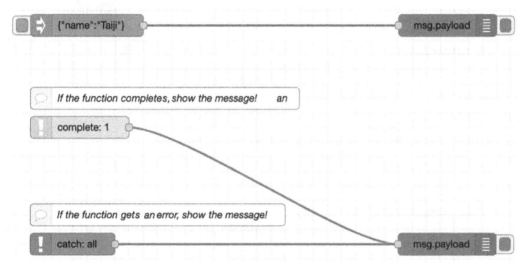

Figure 4.3 – The flow with our common category nodes

The data in the **inject** node is simple JSON data here. Double-click the placed **inject** node to open the settings panel and set the JSON data. Please refer to the following:

```
{"name":"Taiji"}
```

You can change the JSON data in the **inject** node for what you want to send. Also, you should set the properties for the **complete** node. Open the settings panel and set a node to watch the status.

Set each node's parameters as follows:

- The **inject** node:

  Please set the first parameter as msg.payload with the following JSON:

  ```
  {"name": "Taiji"}
  ```

  You can set any value here:

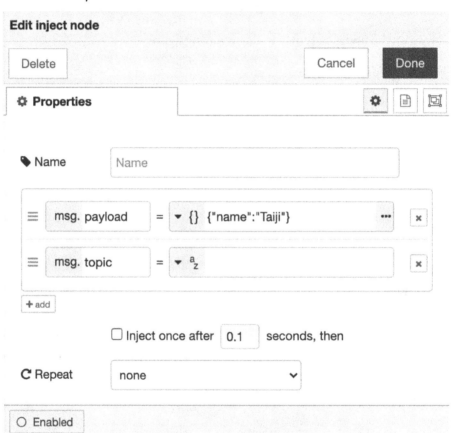

Figure 4.4 – An inject node for inserting data

- The **complete** node:

  Check the first option of the **Properties** tab to watch the status of the **inject** node:

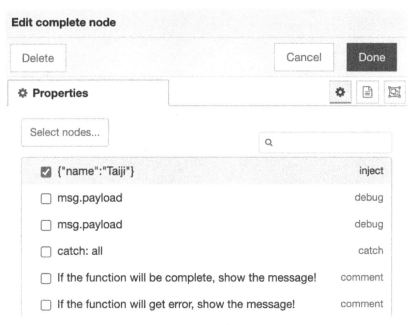

Figure 4.5 – A complete node for watching the status

No properties of other nodes need to be changed.

After the setting changes, you need to deploy and click the button of the **inject** node. After that, you can see the JSON data in the right-hand panel of the **debug** tab.

You can get the flow definition from the book's GitHub repo at `https://github.com/PacktPublishing/-Practical-Node-RED-Programming/blob/master/Chapter04/common-flows.json`.

Function category

In this section, we will learn how to use some major nodes from the function category, and will make a flow with these nodes.

Create a sample flow using the nodes in the function category. Here, we will use the following six nodes:

- The **inject** node
- The **function** node
- The **switch** node
- The **change** node
- The **template** node

- The **debug** node

Place and wire the nodes as shown in the following figure:

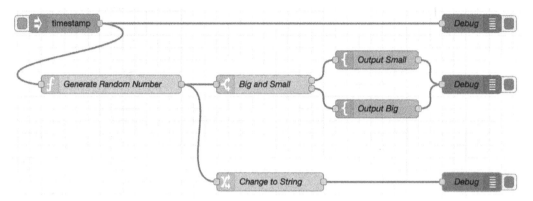

Figure 4.6 – The flow with function category nodes

Please follow these steps to make the flow:

1. Place the **inject** node and **debug** node on the palette. These two nodes can be used with their default parameters. No change of settings is required here.

2. Place a **function** node on the palette.

3. Open the settings panel of the **function** node and enter the following code:

```
// generate random number
var min = 1 ;
var max = 10 ;
var a = Math.floor( Math.random() * (max + 1 - min) ) +
   min ;

// set random number to message
msg.payload = a;

// return message
return msg;
```

4. After coding, click on **Done** to save the settings:

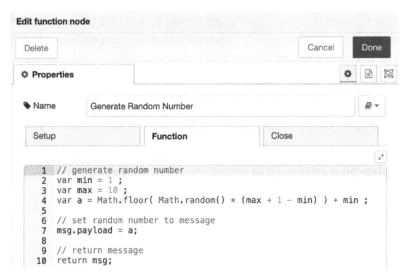

Figure 4.7 – Function node settings

5. Place the **switch** node on the palette, then open the settings panel of the **switch** node and set the value rules as follows:

- The < field: 6
- The > field: 5

This should look as follows:

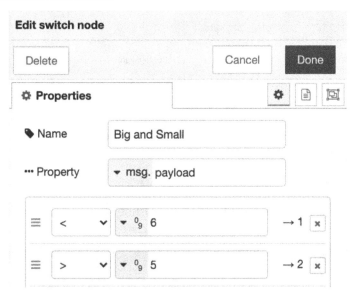

Figure 4.8 – The switch node settings

If the input parameter is 5 or less, the output route is 1, and if the input parameter is 6 or more, the output route is 2. This means that the next node depends on the number of input parameters.

6. Place two **template** nodes on the palette.

    The previous function was the **switch** node, so the data splits depending on the result of the output.

7. Open the settings panel of each **template** node and enter the following code for the first **template** node connected to output route 1 of the **switch** node:

```
The number is small: {{payload}} !
```

The **template** node will look something like the following screenshot once we add the preceding code:

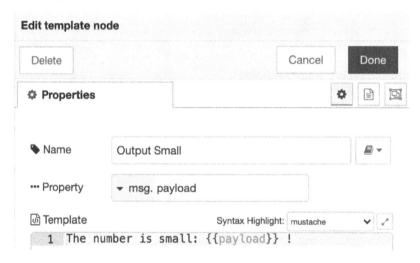

Figure 4.9 – The first template node settings

8. Enter the following code for the second **template** node, which is connected to output route 2 of the **switch** node:

```
The number is big: {{payload}} !
```

It will look something like the following screenshot:

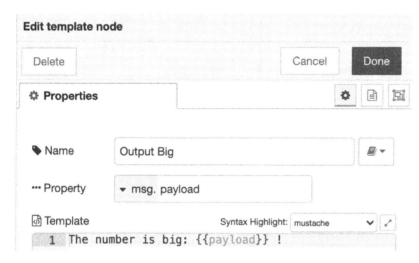

Figure 4.10 – The second template node settings

9. Place the **change** node on the palette, open the settings panel of the **change** node, and look at the settings box below **Rules**.

10. Select **string** from the drop-down menu in the box next to **to** and enter the desired character string in the text box next to this. Here, it says **It has been changed to string data!**. Please refer to the following screenshot:

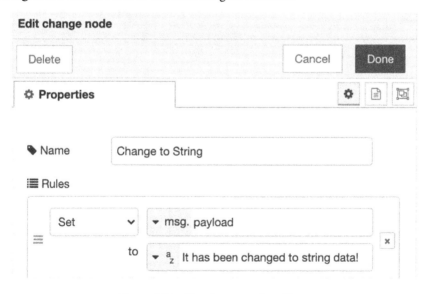

Figure 4.11 – The change node settings

11. After changing the settings, you need to deploy and click the button of the **inject** node.

Once you do this, you can see the data in the debug tab in the right-hand panel, as follows:

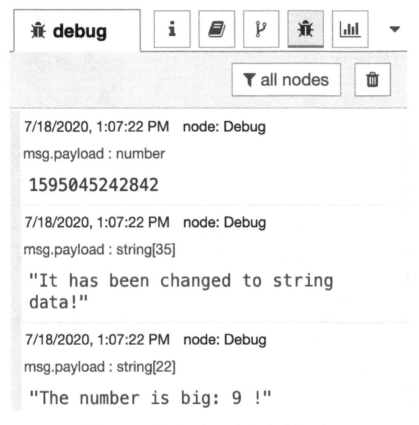

Figure 4.12 – Showing the results in the debug tab

The first debug message is the default **inject** node value as a timestamp. The second one is the debug message of the **debug** node placed after the **change** node. The last one depends on the random number and is formatted by the **template** node.

You can get the flow definition from the book's GitHub repo at `https://github.com/PacktPublishing/-Practical-Node-RED-Programming/blob/master/Chapter04/function-flows.json`.

Next, let's learn about nodes that are not provided by default.

Getting several nodes from the library

You can get several more attractive nodes that have been developed by Node-RED contributors and install them in your Node-RED flow editor. You can find new nodes, share your flows, and see what other people have done with Node-RED. In this section, we will learn how to get several other nodes from the Node-RED library. Let's first access the Node-RED library site: `https://flows.nodered.org/`. In the following screenshot, you can see how the Node-RED library looks:

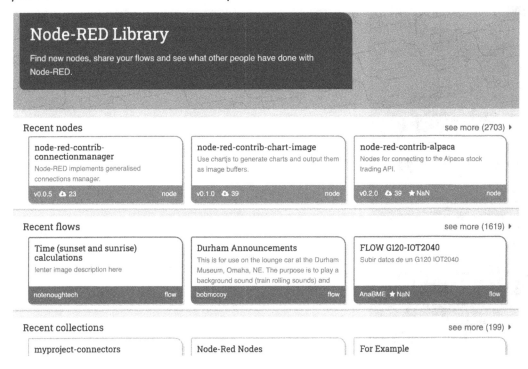

Figure 4.13 – Node-RED Library

It's easy to use this library in your own Node-RED environment's flow editor. Let's see how to install a node from the library:

1. Select **Manage palette** from the sidebar menu. You will see the **User Settings** panel open with the **Palette** tab selected.

2. Type watson in the search field, or the name of any other node you want to use. If you find the node you want, click the **Install** button:

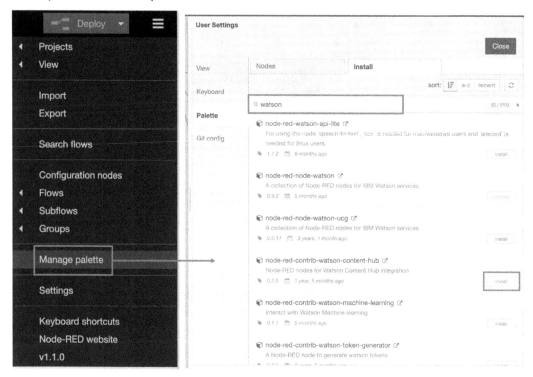

Figure 4.14 – Opening the User Settings panel and finding the node you want to use

3. After clicking on the **Install** button, a pop-up window will appear, on which you will need to click on **Install** once again.

    Once you do this and the installation has completed, you will get a pop-up message saying **Nodes added to palette**.

That's all! You can see all the nodes you have installed in your palette as shown in the following figure:

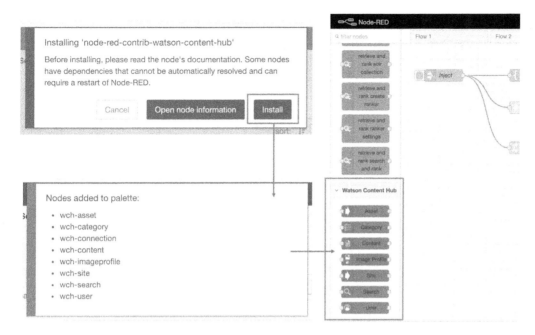

Figure 4.15 – Nodes you have installed are added to your palette

> **Tip**
>
> You can search for useful nodes on the Node-RED Library website. It's possible
> to search by keywords, and sort the results in terms of most recently added,
> number of downloads, and ratings. I recommend sorting by number of
> downloads first because nodes that have been downloaded by lots of developers
> are likely to be very useful: `https://flows.nodered.org/`
> `search?type=node&sort=downloads`.

Now you have become a great Node-RED user and have mastered how to use the Node-
RED flow editor to make some flows (applications).

Summary

In this chapter, you've learned how to use each major node in the Node-RED flow editor.
You have successfully made your Node-RED flows! The flow steps you've created here are
most of the steps you will need to do to create various flows in the future.

The important point learned in this chapter is that each node has its own unique features. By combining these like a puzzle, we can create an application similar to one made through regular programming just by creating a flow.

In the next chapter, let's create a more practical sample flow (application) for IoT edge devices.

Section 2:
Mastering Node-RED

In this section, readers will actually create an application using the Node-RED flow editor. Instead of trying to build advanced applications from the beginning, first they will learn how to create a sample flow for each major environment (that is, stand-alone environments such as the Raspberry Pi, desktop, and cloud).

In this section, we will cover the following chapters:

- *Chapter 5, Implementing Node-RED Locally*
- *Chapter 6, Implementing Node-RED in the Cloud*
- *Chapter 7, Calling a Web API from Node-RED*
- *Chapter 8, Using the Project Feature with Git*

5
Implementing Node-RED Locally

In this chapter, let's use the standalone version of Node-RED. Node-RED consists of a development environment, an execution environment, and the application itself. You can understand the mechanism by using the standalone version that runs in the local environment.

Specifically, the most common reason for starting the standalone version of Node-RED is when using it on an IoT edge device. IoT edge devices have sensors that are usually applied to the "Things" part of the "Internet of Things." In this chapter, we will look at the sensing data within the edge device and create a sample flow.

Let's get started with the following four topics:

- Running Node-RED on a local machine
- Using the standalone version of Node-RED
- Using IoT on edge devices
- Making a sample flow

By the end of this chapter, you will have learned how to build a flow for handling sensor data on IoT devices.

Technical requirements

To progress through this chapter, you will need the following:

- Node-RED (v1.1.0 or above): `https://nodered.org/`
- Raspberry Pi: `https://www.raspberrypi.org/`

The code used in this chapter can be found in `Chapter05` folder at `https://github.com/PacktPublishing/-Practical-Node-RED-Programming`.

Running Node-RED on a local machine

We can now create the flow for sensing data on an IoT edge device, and in this scenario, the local machine uses Raspberry Pi. The reason for this will be described in the *Using the standalone verison of Node-RED* section, but in summary, this tutorial is for IoT edge device.

I have already explained how to start Node-RED on Raspberry Pi, so you should now know how to run it, but if you need a refresher, please refer to the *Install Node-RED for Raspberry Pi* section in *Chapter 2*, *Setting Up the Development Environment*.

Now, follow these steps to start Node-RED on your Raspberry Pi:

1. Let's start by executing Node-RED from the Raspberry Pi menu:

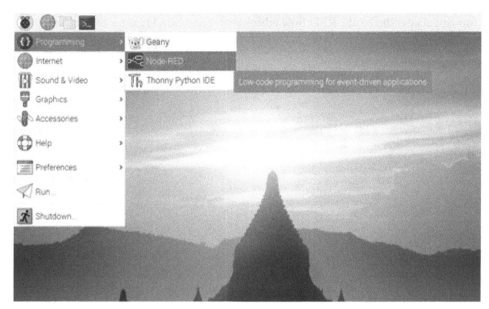

Figure 5.1 – Running Node-RED from the Raspberry Pi menu

2. You can check the status of Node-RED on your terminal. If **Started flows** is shown, Node-RED is ready to use:

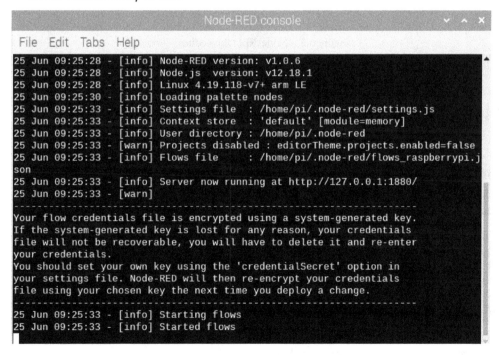

Figure 5.2 – Terminal of Raspberry Pi

3.   You can access the Node-RED flow editor with the `localhost:1880` URL:

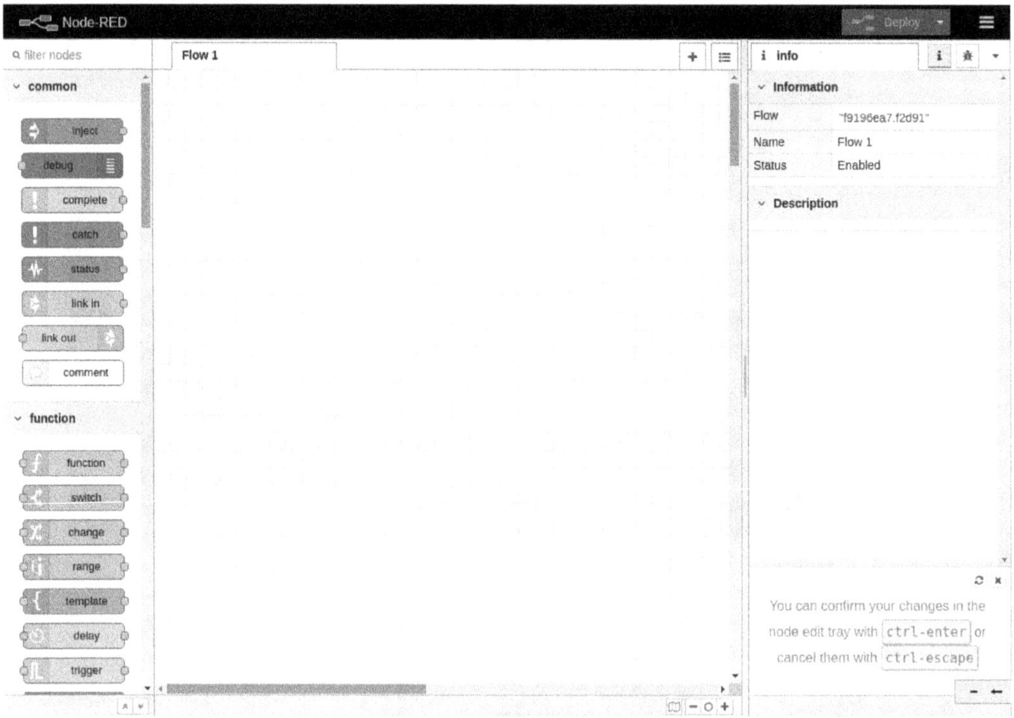

Figure 5.3 – Node-RED flow editor

Let's learn a few concepts before making use of the flow editor.

Using the standalone version of Node-RED

Now we will learn what the standalone version of Node-RED is and how it differs from other versions. We usually use the Node-RED flow editor as a standalone editor; however, we can also use the Node-RED flow editor on any cloud with container technologies such as Docker, Kubernetes, or Cloud Foundry. We will explicitly demonstrate the use of the standalone version with relatively common use cases to learn how to use it.

Let's think about situations where Node-RED is used.

Node-RED is a tool for creating applications made with Node.js. It is also the execution environment. If you can write an application in Node.js, that's fine.

So, why build an application with Node-RED?

One answer is to black-box each individual unit of data processing. This makes the role of each process very clear and easy to build and maintain.

Another answer is to avoid human error. Since each process is modularized as a node, you only need to understand the input/output specifications when using that process. This means you can avoid human errors such as coding mistakes and missing test specifications. This can be the advantage of no-code/low-code as well as Node-RED.

Next, imagine a concrete situation that uses Node-RED with the characteristics just described.

Think of a business logic that controls data and connects it to the next process. This is a common situation in IoT solutions.

The standard architecture for IoT solutions is built with edge devices and cloud platforms. It sends the sensor data acquired by the edge device to the cloud and then, on the cloud work to process the data, such as visualizing, analyzing, and persistent.

In this chapter, I would like to focus on that edge device part.

It is common for edge devices to want to prepare the acquired sensor data to some extent before sending it to the cloud. The reason for this that if you send all the acquired data, there is a risk that the network will be overloaded.

So, the standalone Node-RED exercise uses Raspberry Pi, which is a famous IoT infrastructure for consumers.

In this chapter, we will use the **Grove Base HAT** for Raspberry Pi and Grove Base modules. This is one of the standards for the IoT edge device platform and so we need to install the Grove Base driver to Raspberry Pi.

> **Important Note**
>
> This chapter gives an example using Grove Base HAT, which is relatively inexpensive and can be purchased (the link to this is mentioned in the next section), but any sensor device that can be connected to a Raspberry Pi can handle data on Node-RED.
>
> When using a module other than the Grove Base HAT sensor device, use the corresponding node and read this chapter. (Implementation is required if there is no corresponding node.)
>
> You can check the Node-RED library for the existence of a node that corresponds to each device:
>
> ```
> https://flows.nodered.org/
> ```

Let's prepare to use Grove Base HAT on our Raspberry Pi by following these steps:

1. Let's start by executing the following command on our Raspberry Pi:

    ```
    $ curl -sL https://github.com/Seeed-Studio/grove.py/raw/
    master/install.sh | sudo bash -s -
    ```

2. If everything goes well, you will see the following notice:

```
File Edit Tabs Help
Requirement already satisfied: smbus2 in /usr/local/lib/python3.7/dist-packages (from sgp30) (0.3.0)
Installing collected packages: sgp30
Successfully installed sgp30-0.1.6
Looking in indexes: https://pypi.org/simple, https://www.piwheels.org/simple
Collecting https://github.com/Seeed-Studio/grove.py/archive/master.zip
  Downloading https://github.com/Seeed-Studio/grove.py/archive/master.zip
     / 696kB 13.5MB/s
Requirement already satisfied, skipping upgrade: RPi.GPIO in /usr/lib/python2.7/dist-packages (from grove.py==0.6) (0.7.0)
Requirement already satisfied, skipping upgrade: rpi_ws281x in /usr/local/lib/python2.7/dist-packages (from grove.py==0.6) (4.2.3)
Requirement already satisfied, skipping upgrade: smbus2 in /usr/local/lib/python2.7/dist-packages (from grove.py==0.6) (0.3.0)
Building wheels for collected packages: grove.py
  Running setup.py bdist_wheel for grove.py ... done
  Stored in directory: /tmp/pip-ephem-wheel-cache-QrvP_1/wheels/1a/5b/70/01a949561c39a7059cd1daae9fa6d03e7b2c58d7ec4fb6245f
Successfully built grove.py
Installing collected packages: grove.py
Successfully installed grove.py-0.6
Looking in indexes: https://pypi.org/simple, https://www.piwheels.org/simple
Collecting https://github.com/Seeed-Studio/grove.py/archive/master.zip
  Using cached https://github.com/Seeed-Studio/grove.py/archive/master.zip
Requirement already satisfied, skipping upgrade: RPi.GPIO in /usr/lib/python3/dist-packages (from grove.py==0.6) (0.7.0)
Requirement already satisfied, skipping upgrade: rpi_ws281x in /usr/local/lib/python3.7/dist-packages (from grove.py==0.6) (4.2.3)
Requirement already satisfied, skipping upgrade: smbus2 in /usr/local/lib/python3.7/dist-packages (from grove.py==0.6) (0.3.0)
Building wheels for collected packages: grove.py
  Running setup.py bdist_wheel for grove.py ... done
  Stored in directory: /tmp/pip-ephem-wheel-cache-bhetx0r8/wheels/1a/5b/70/01a949561c39a7059cd1daae9fa6d03e7b2c58d7ec4fb6245f
Successfully built grove.py
Installing collected packages: grove.py
Successfully installed grove.py-0.6
#######################################################
  Lastest Grove.py from github install complete   !!!!!
#######################################################
pi@raspberrypi:~ $
```

Figure 5.4 – Successful grove.py installation

3. The next step is to enable ARM I2C. We can do this by executing the following command:

    ```
    $ sudo raspi-config
    ```

4. After executing the command, you will see the following configuration window. Please select **Interfacing Options**:

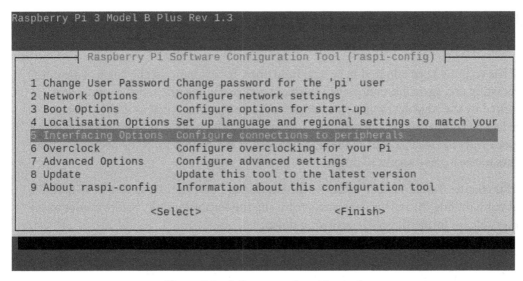

Figure 5.5 – Software configuration tool

5. Select **I2C**:

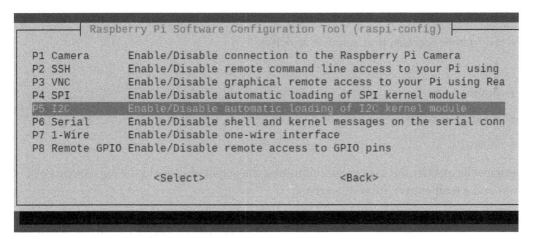

Figure 5.6 – Enabling I2C

6. Once you select it, a **Would you like the ARM I2C interface to be enabled?** message will be shown in the same window. Please select **Yes** to accept it.

You have now successfully enabled I2C. Restart the Raspberry Pi and restart the Node-RED flow editor. In doing this, your Raspberry Pi has been made available to use the I2C interface, and for the next step, we need to connect the sensor devices and Raspberry Pi via the I2C interface.

Using IoT on edge devices

Now let's consider a case study on edge devices in IoT.

IoT has recently been adopted in several industries, for example, in the fields of weather forecasting and agriculture; however, the basic composition is the same. Various data acquired by the edge device is sent to the server-side platform, such as the cloud, and the data is handled and visualized on the server side, which is full of resources. There are various ways to visualize, but in the simplest case, it will be to output the necessary data values to the log as a standard output.

In this chapter, I would like to consider the edge device part in the use case of IoT. This is about handling the sensor data, acquired using the sensor module, before it goes to the server side for formatting and narrowing down.

What are the different kinds of sensors?

The following sensors are often used at the experimental level of IoT:

- Temperature
- Humidity
- Gyroscope (acceleration, angular velocity)
- Light
- Sound
- Pressure-sensitive
- Magnetic

Here we will consider the use case of outputting the acquired value to the log using a light sensor and a temperature/humidity sensor.

In order to get sensor data, you'll need a device. In this sample flow (application), Raspberry Pi is used, but it does not have a sensing function because it is just a foundation. With the old-fashioned board, you had to solder the sensor device/module, but the convenient thing about the Raspberry Pi is that there are many sensor module kits that can be connected with one touch.

As already introduced, we'll use the Grove series provided by Seeed, which has a sensor module and connection board for Raspberry Pi: `https://wiki.seeedstudio.com/Grove_Base_Hat_for_Raspberry_Pi/`

Let's prepare the Grove Base HAT for Raspberry Pi modules.

> **Important Note**
> If you don't have the Grove Base HAT for Raspberry Pi and want to run this
> tutorial, please buy it via the official site (`https://www.seeedstudio.`
> `com/Grove-Base-Hat-for-Raspberry-Pi.html`).

This is what the Grove Base HAT for Raspberry Pi looks like:

Figure 5.7 – Grove Base HAT for Raspberry Pi

We need to connect the Grove Base HAT and the sensor modules to the Raspberry Pi. To
do so, follow these steps:

1. Place the Grove Base HAT on your Raspberry Pi and screw it in:

Figure 5.8 – Setting the Base HAT on your Raspberry Pi

This is what the Grove - Light Sensor v1.2 - LS06-S phototransistor looks like:

Figure 5.9 – Grove - Light Sensor v1.2

You can get it from `https://www.seeedstudio.com/Grove-Light-Sensor-v1-2-LS06-S-phototransistor.html`.

2. Connect the Grove light sensor to the analog port of your Base HAT:

Figure 5.10 – Connecting the light sensor to your Base HAT

> **Important Note**
>
> Please be careful! This vendor, **Seeed,** has a similar module for temperature/ humidity sensor **SHT35**, but it's not supported by the Grove Base HAT node. You need to use **SHT31**.

This is what the Grove - Temperature&Humidity Sensor (SHT31) looks like:

Figure 5.11 – Grove – Temperature&Humidity Sensor (SHT31)

You can get it from `https://www.seeedstudio.com/Grove-Temperature-Humidity-Sensor-SHT31.html`.

3. Connect the Grove temperature and humidity sensor to the I2C port of your Base HAT:

Figure 5.12 – Connecting the temperature/humidity sensor to your Base HAT

And that's it. Now your device is set up and we are ready to go on to the next step! In this part, we have learned about popular, simple use cases of IoT edge devices and next, we will make a flow for these use cases.

Making a sample flow

In this section, we will create these two sensor data output flows in the Node-RED flow editor.

You will use the sensor modules you have prepared to collect data and create a sample flow to visualize it on Node-RED. By using two different sensor modules, we can learn the basics of data handling in Node-RED.

Use case 1 – light sensor

The first is a light sensor. Let's create a flow (application) that detects light and outputs the value detected by a fixed-point observation to a log:

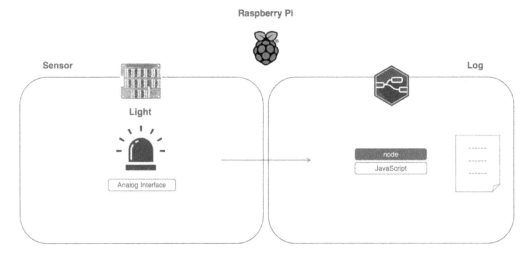

Figure 5.13 – Use case 1 – getting light sensor data

Connect the light sensor module to the Raspberry Pi and use the Node-RED flow editor on the Raspberry Pi to output the data obtained as a standard output.

Use case 2 – temperature/humidity sensor

The second one is a temperature/humidity sensor. Let's create an application (flow) that detects temperature and humidity and outputs the value detected by a fixed-point observation to a log:

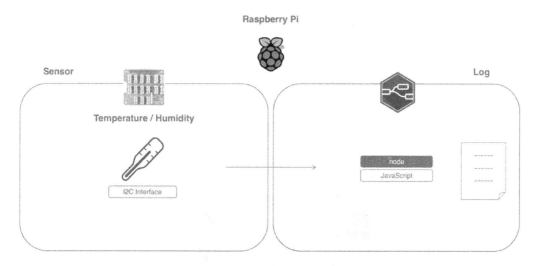

Figure 5.14 – Use case 2 – getting temperature/humidity data

Connect the temperature/humidity sensor module to the Raspberry Pi and use the Node-RED flow editor on the Raspberry Pi to output the data obtained as a standard output.

If you want to spot test these two use cases on your device, you need to connect a sensor that you can use to obtain sensor data.

You may have to prepare this before creating the flow.

This time, we will use Grove Base HAT, which is easy to use with Raspberry Pi, and as this setup was completed in the previous step, we are ready to access the data on Raspberry Pi. However, we have not yet prepared Node-RED. It is difficult to access this data with Node-RED as default. One way is to use a Function node and code the script from scratch, which is very difficult but not impossible.

For handling the sensing data recognized by Raspberry Pi on Node-RED, a "node" dedicated to Grove Base HAT is required.

The good news is that you can start using the node right away. This is because Seigo Tanaka, a Node-RED User Group Japan board member (https://nodered.jp/) and Node-RED contributor, has already created and released a node for Grove Base HAT. This is the node for the Grove Base HAT for Raspberry Pi:

```
node-red-contrib-grove-base-hat
```

You can read more about it here: https://www.npmjs.com/package/node-red-contrib-grove-base-hat.

If you need a refresher on how to install nodes that are published on the node library, please read the *Getting several nodes from the library* section in *Chapter 4, Learning the Major Nodes*.

The reason I refer you back to this is that the next step is to install the node for the Grove Base HAT from the library into your environment.

Let's enable the use of this Grove Base HAT node in our Node-RED flow editor:

1. Click the menu at the top right and select **Manage palette** to open the settings panel:

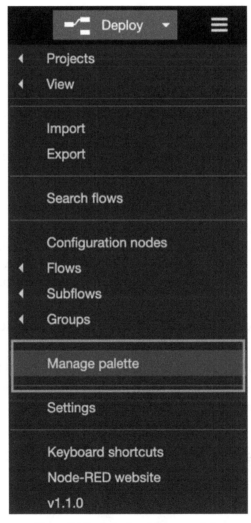

Figure 5.15 – Selecting Manage palette

2. When the settings panel is opened, type the name of the node you want to use in the search window. We want to use **node-red-contrib-grove-base-hat**, so please type the following:

```
grove base
```

3. After that, you can see the **node-red-contrib-grove-base-hat** node in the search window. Click the **Install** button:

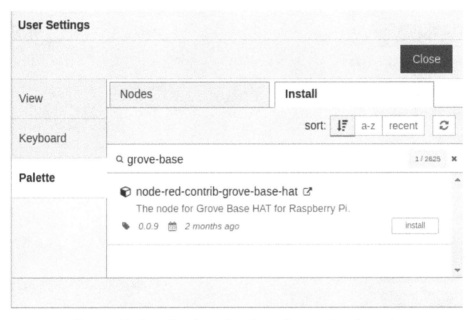

Figure 5.16 – Installing the node-red-contrib-grove-base-hat node

4. After clicking the **Install** button, you will see a message asking you to read the documentation to find out more information about this node. Read the document if necessary, and then click the **Install** button on the message box:

Installing 'node-red-contrib-grove-base-hat'

Before installing, please read the node's documentation. Some nodes have dependencies that cannot be automatically resolved and can require a restart of Node-RED.

Figure 5.17 – A message window to read the node documentation

Now you are ready to use the node for Grove Base HAT. Check the palette in the flow editor. At the bottom of the palette, you can see that the Grove Base HAT node has been added:

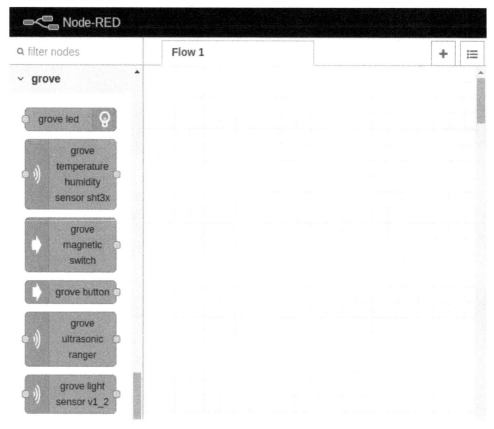

Figure 5.18 – Grove Base HAT nodes on your dashboard

There are many sensing modules that can be connected to Grove Base HAT. This time, only the light and temperature/humidity sensors are used, but there are other things that can be seen by looking at the types of nodes.

The procedure followed for the two use cases created here can also be applied when using other sensors. If you are interested, please try other sensors too. In the next section, we will make a flow for use case 1.

Making a flow for use case 1 – light sensor

In use case 1, Node-RED can be used to handle the illuminance obtained from the light sensor as JSON data. That data can be handled as JSON data, then be sent to the server side afterward, and various processes can be easily performed on the edge device.

The value obtained from the light sensor is received by Node-RED and the output is a debug log (standard output). We can set this using the following steps:

1. Select the **grove light sensor v1_2** node from the palette on the left side of the flow editor and drag and drop it into the workspace to place it:

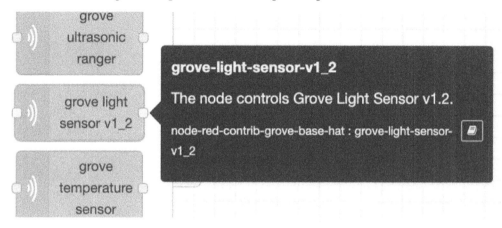

Figure 5.19 – grove light sensor v1_2

This node allows the value of the sensor device, which is continuously acquired on the Raspberry Pi via the Grove Base HAT, to be handled as a JSON format message object on Node-RED.

2. After placing the **grove-light-sensor-v1_2** node, place the **inject** node and **debug** nodes and wire them so that the **grove-light-sensor-v1_2** node you placed is sandwiched between them:

Figure 5.20 – Placing nodes and wiring them for the light sensor

3. Next, check the settings of the **grove-light-sensor-v1_2** node. Double-click the node to open the settings panel.

4. There is a selection item called **Port** in the settings panel. **A0** is selected by default.

    This **Port** setting is to specify which connector on the Grove Base HAT gets data from the connected module.

5. Earlier, we connected the Grove light sensor to the Grove Base HAT. If the connection is made according to the procedure in this tutorial, it should be connected to port A2, so select **A2** as the node setting value. If you are connecting to another port, select the port you are connecting to:

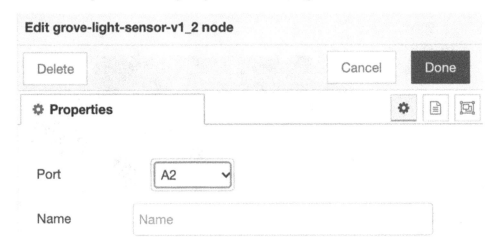

Figure 5.21 – Select A2 as the port if you connected the sensor to A2 of Base HAT

6. After checking and setting **Port** on the settings panel, click the **Done** button in the upper-right corner to close the settings panel.

That's it! Don't forget to click the **deploy** button.

You should remember how to execute a flow from a inject node, because you learned about this in the previous chapter. Click the switch on the inject node to run the flow. The data for the timing when the switch is clicked is outputted as a log, so please try clicking it a couple of times.

Important Note

Do not forget to display the debug window to show that the value of the acquired data will be the output to the debug window. Node-RED does not automatically show the debug window even if the debug output is activated.

The resulting output in the **debug** window looks like the following:

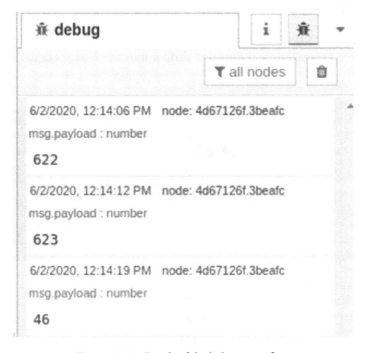

Figure 5.22 – Result of the light sensor flow

You can see that the result was output to the **debug** window.

Congratulations! With this, we have successfully created a basic flow (application) that handles the value of our first light sensor with Node-RED.

You can also download this flow definition file here: `https://github.com/ PacktPublishing/-Practical-Node-RED-Programming/blob/master/ Chapter05/light-sensor-flows.json`.

Making a flow for use case 2 – temperature/humidity sensor

In use case 2, Node-RED can be used to handle the temperature and the humidity obtained from the temperature/humidity sensor as JSON data. The data, which can be handled as JSON data, can be sent to the server side afterward, and various processes can be easily performed on the edge device.

The value obtained from the temperature/humidity sensor is received by Node-RED and is outputted as a debug log (standard output):

1. Select the **grove temperature humidity sensor sht3x** node from the palette on the left side of the flow editor and drag and drop it into the workspace to place it:

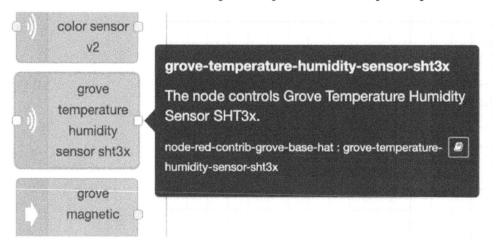

Figure 5.23 – grove temperature humidity sensor sht3x

This node allows the value of the sensor device, which is continuously acquired on the Raspberry Pi via Grove Base HAT, to be handled as a JSON format message object on Node-RED.

2. After placing the **grove-temperature-humidity-sensor-sht3x** node, place the **inject** and **debug** nodes, respectively, and wire them so that the **grove-temperature-humidity-sensor-sht3x** node you placed is sandwiched between them:

Figure 5.24 – Placing the nodes and wiring them for the temperature and humidity sensor

3. Next, check the settings of the **grove-temperature-humidity-sensor-sht3x** node and double-click the node to open the settings panel.

Actually, this node has no values to set (strictly speaking, the name can be set, but the presence or absence of this setting does not affect the operation):

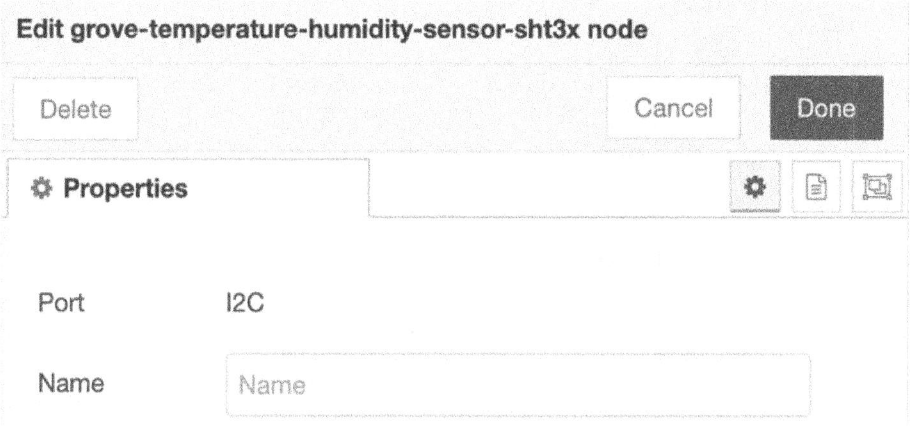

Edit grove-temperature-humidity-sensor-sht3x node

| Delete | | | Cancel | Done |

| ⚙ **Properties** | | | ⚙ | 📄 | 🔲 |

| Port | I2C |
| Name | Name |

Figure 5.25 – Already set to the I2C port

You can see on the settings panel that the port is designated as **I2C** (not changeable). If you have connected the Grove temperature and humidity sensor to the Grove Base HAT according to the procedure in this document, the module should be correctly connected to the **I2C** port. If it is connected to a port other than I2C, reconnect it properly.

4. After checking **Port** on the settings panel, click the **Done** button in the upper-right corner to close the settings panel.

    That's it! Don't forget to click the **deploy** button.

5. Click the switch on the inject node to run the flow. The data for the timing when the switch is clicked is outputted as a log, so please try clicking it a couple of times.

> **Important Note**
>
> As noted in the previous section, do not forget to display the debug window to show that the value of the acquired data will be the output to the debug window. Node-RED does not automatically show the debug window even if the debug output is activated.

The resulting output in the **debug** window looks like the following:

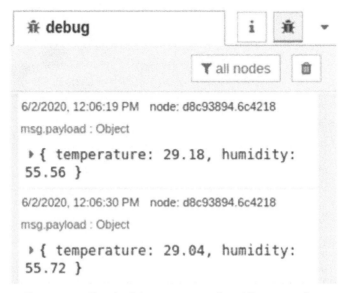

Figure 5.26 – Result of the temperature/humidity sensor flow

You can see that the result was outputted to the **debug** window.

Congratulations! With this, we have successfully created a basic flow (application) that handles the value of the second sample, the temperature/humidity sensor, with Node-RED.

You can also download this flow definition file here: https://github.com/ PacktPublishing/-Practical-Node-RED-Programming/blob/master/ Chapter05/light-sensor-flows.json.

Well done! Now you have learned how to handle the data obtained from the illuminance sensor and temperature and humidity sensor in JSON format on Node-RED.

Summary

In this chapter, you learned how to create a sample flow (application) by comparing Node-RED to a real IoT use case. We experienced using the sensor module and Raspberry Pi to exchange data with Node-RED, so we had a feel for IoT.

The flow steps created here will help you create different flows with other sensor modules in the edge device in the future.

In the next chapter, we will use the IoT use case as we did this time, but we will create a practical sample flow (application) on the cloud side (server side).

6
Implementing Node-RED in the Cloud

In this chapter, we will learn how to utilize Node-RED, which can be used standalone on a cloud platform (mainly Platform as a Service). **Platform as a Service (PaaS)** provides an instance that acts as the execution environment for an application, and the application developers only focus on executing the application created by themselves without using their power to build the environment. Node-RED is actually a Node.js application, so you can run it wherever you have a runtime environment for Node.js.

There are various major mega clouds such as Azure, AWS, and GCP, but Node-RED is prepared as a Starter App (a web application that can be launched on IBM Cloud is called a Starter App) by default in IBM Cloud, so we will use it in this chapter.

In this chapter, we'll cover the following topics:

- Running Node-RED on the cloud
- What is the specific situation for using Node-RED in the cloud?
- IoT case study spot on the server side
- Making a sample flow

By the end of this chapter, you will have mastered how to build a flow for handling sensor data on the cloud.

Technical requirements

The code that will be used in this chapter can be found in the `Chapter06` folder at `https://github.com/PacktPublishing/-Practical-Node-RED-Programming`.

Running Node-RED on the cloud

This time, we will use IBM Cloud. The reason for this is that IBM Cloud has Node-RED Starter Kit on it. This is a kind of software boilerplate that includes services needed for Node-RED on the cloud, such as a database, CI/CD tools, and more.

If you have not used IBM Cloud yet, don't worry – IBM provides a free IBM Cloud account (Lite account) with no credit card registration needed. You can register for an IBM Cloud Lite account at `http://ibm.biz/packt-nodered`.

Before using Node-RED on IBM Cloud, you need to finish the registration process for your IBM Cloud Lite account.

> **Important Note**
>
> In this book, we strongly recommend that you select a Lite account when using IBM Cloud. You can upgrade from a Lite account to a standard account (PAYG/Pay as you go) at your own will. This means you can automatically upgrade to PAYG by registering your credit card.
>
> Please note that services that can be used free of charge with a Lite account may be charged for with PAYG.

Now, let's launch Node-RED on IBM Cloud by following these steps:

> **Important Note**
>
> The instructions/screenshots provided here are correct at the time of writing. The UI of IBM Cloud changes so often that it might be different from the current UI.

1. Log in to IBM Cloud (`https://cloud.ibm.com`) with the account you created previously:

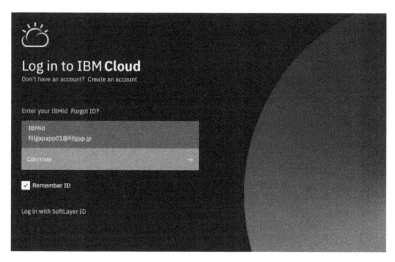

Figure 6.1 – Logging in via your Lite account

2. After logging into IBM Cloud, you will see your own dashboard on your screen. If this is your first time using IBM Cloud, no resources will be shown on the dashboard:

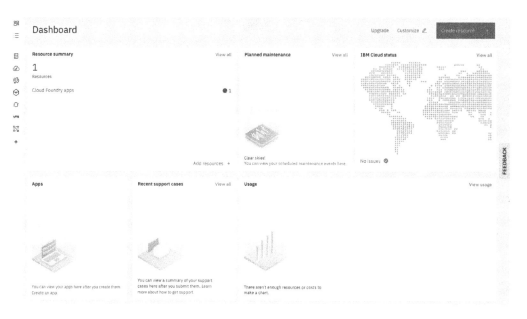

Figure 6.2 – IBM Cloud dashboard

Next, we will create Node-RED on this cloud platform.

3. We will create Node-RED as a service on this cloud. Click **App Development** from the menu at the top left and click the **Get a Starter Kit** button. This lets you create a new application service:

Figure 6.3 – Get a Starter Kit button

4. You can find Node-RED if you type Node-RED into the search text box. Once you've found it, click on the **Node-RED** panel:

Figure 6.4 – Node-RED Starter Kit

5. After clicking on the **Node-RED** panel, we need to set some items.

You can freely change each item by providing your own values, but in this chapter, the values that have been set here will be used for explanation purposes.

See *Figure 6.5* for the settings and values to configure. Please note that once they are set, these items cannot be changed later.

6. After setting all the items, click the **Create** button:

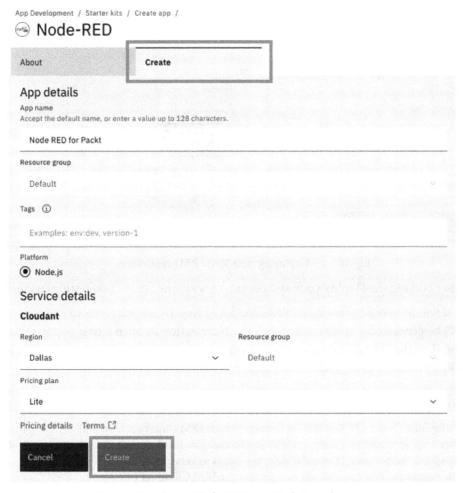

Figure 6.5 – Create Node-RED as a Node.js application

You have now created the framework for the applications that make up Node-RED. After this, you will be redirected to the **App Details** screen automatically, where you will be able to see that the **Cloudant** instance of the linked service has also been provisioned.

However, only the application source code and the instance of the cooperation service are created, and they haven't been deployed to the Node.js execution environment on IBM Cloud yet. The actual deployment will be done when the CI/CD toolchain is enabled.

7. When everything is ready, click on the **Deploy your app** button in the center of the screen to enable it:

Figure 6.6 – Deploying your Node-RED application

8. After clicking the **Deploy your app** button, move to the application settings window.

9. You will be asked to create an IBM Cloud API Key. Don't worry about this, as one will be generated automatically. Click the **New** button to open a new popup window, and then the **OK** button on the popup window. Once you do this, an IBM Cloud API Key will be generated:

> **IBM Cloud API Key**
>
> The IBM Cloud API Key is used to control your IBM Cloud account and various services (for example, it's Cloud Foundry in this tutorial). You can use this to issue a token for external access to services on IBM Cloud, for example. You can find out more about the IBM Cloud API Key here: `https://cloud.ibm.com/docs/account?topic=account-manapikey`.

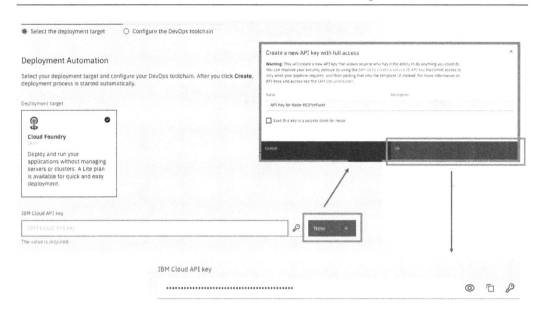

Figure 6.7 – Generating an IBM Cloud API Key

10. Select the resource spec on the window.

    This time, we are using IBM Cloud with a Lite account, so we have only 256 MB of memory available for all our services on IBM Cloud. So, if we use 256 MB for the Cloud Foundry Node.js service, we won't be able to use more memory for other services. But Node-RED needs 256 MB to run on IBM Cloud, so please use 256 MB here. It is already allocated 256 MB for the instance by default, so click the **Next** button, with no parameters changed:

Figure 6.8 – Node.js runtime instance details

Once you've done this, a **DevOps toolchain** setting screen will be displayed.

11. Click the **Create** button, with the default values filled in.

You can change the DevOps toolchain name to any name you like. This is the name that identifies the toolchain you've created in IBM Cloud:

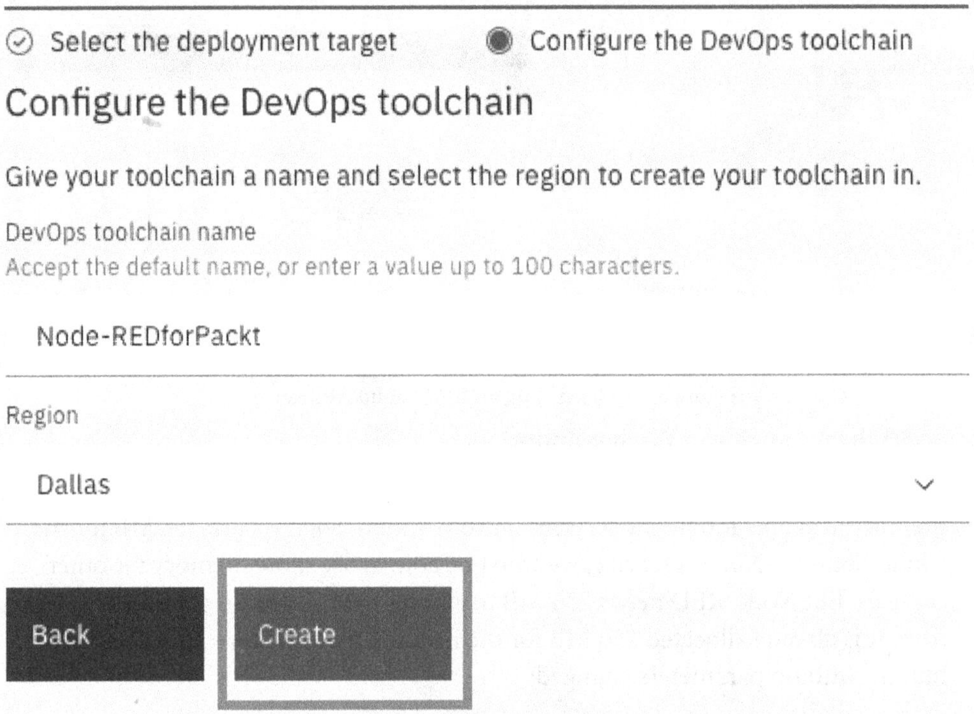

⊘ Select the deployment target   ⦿ Configure the DevOps toolchain

Configure the DevOps toolchain

Give your toolchain a name and select the region to create your toolchain in.

DevOps toolchain name
Accept the default name, or enter a value up to 100 characters.

Node-REDforPackt

Region

Dallas   ⌄

Back   Create

Figure 6.9 – Configure the DevOps toolchain window

Now, you are ready to use the environment (Node.js runtime and DevOps toolchain) to run the Node-RED application you created in the previous step. The Node-RED application you created is automatically deployed on the Node.js runtime through the toolchain.

12. Confirm that the **Status** that's displayed in the **Delivery Pipelines** (pipeline for executing each tool in the DevOps toolchain) area is **Success**, and click the toolchain's name (**Node-REDforPackt**, in this case) above it:

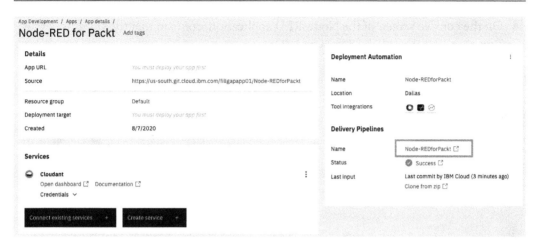

Figure 6.10 – Checking the status of Node-RED and moving to the Pipeline tool

In **Delivery Pipelines**, check that the statuses of both the **BUILD** and **DEPLOY** panels are green and displaying **STAGE PASSED**.

13. Click on **View console** under **LAST EXECUTION RESULT** on the **DEPLOY** panel:

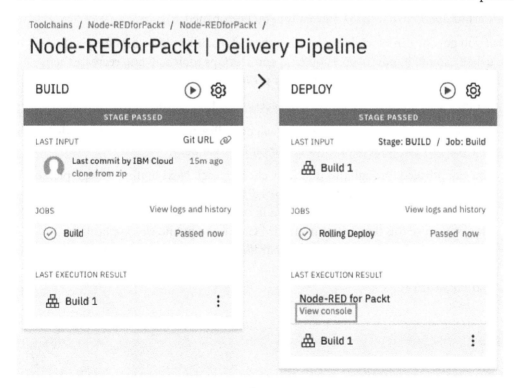

Figure 6.11 – Checking the status of each stage and moving to App Console

14. On the console screen of the Node-RED application, confirm that the status is **Running**, and then click **View App URL**:

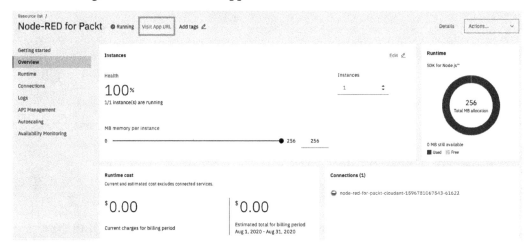

Figure 6.12 – Checking that Node-RED is running and opening Flow Editor

Great work! You opened the Node-RED flow editor on IBM Cloud. Next, we will start to use the Node-RED flow editor you just opened.

If you got any errors while performing these steps, it would be best for you to delete Cloud Foundry App, Cloudant, and DevOps toolchain and recreate them by following the same steps mentioned previously.

15. Set up a **Username** and **Password** to access your flow editor on IBM Cloud.

After clicking on **Visit App URL**, you will be redirected to the initial setup dialog so that you can use Node-RED flow editor on IBM Cloud.

You can proceed through this dialog by clicking each **Next** button, though please note that you should select **Secure your editor so only authorised users can access it** with **Username** and **Password** in order to log in to your own flow editor. This is because this flow editor is on IBM Cloud as a public web application. This means that anybody can access your flow editor if the URL is known. So, I strongly recommend that you select this option and set your own **Username** and **Password** values:

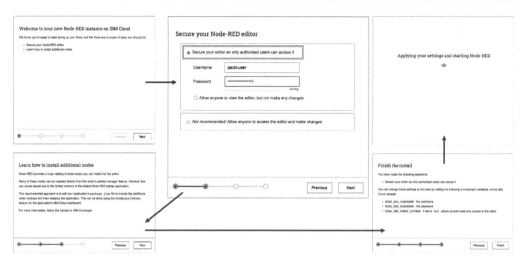

Figure 6.13 – Setting a username and password to access flow editor

We're almost done!

16. Click on the **Go to your Node-RED flow editor** button and then log in with the **Username** and **Password** details that you set in the previous step:

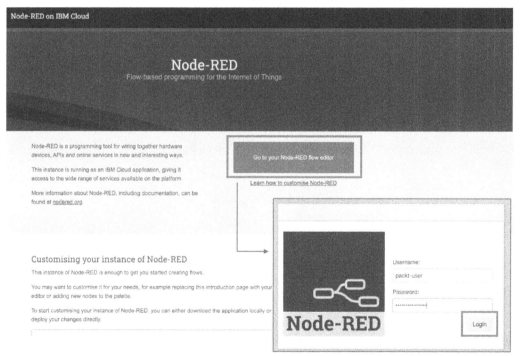

Figure 6.14 – Logging into your Node-RED flow editor

Next, we will check Node-RED flow editor on IBM Cloud and see if it is available.

17. Click the **inject** node and check the result:

Figure 6.15 – Default sample flow

When you click the **inject** node, you will see the resulting value on the **debug** tab:

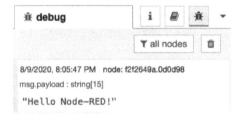

Figure 6.16 – Checking the result

Now, you can create a flow in Node-RED on IBM Cloud. The Node-RED flow editor is always running as a Node.js application on IBM Cloud. This means that the Node.js runtime service (instance) is enabled on IBM Cloud. In other words, unlike Node-RED running on Raspberry Pi, this version of Node-RED accesses the flow editor via the internet.

In the next section, I will explain a little about situations where Node-RED is used on such a cloud.

What is the specific situation for using Node-RED in the cloud?

Let's revisit the situation where Node-RED is used in the cloud.

As we mentioned in the previous chapter, Node-RED is both a tool and an execution environment for creating Node.js applications written in Node.js. As a reason to build an application with Node-RED, I explained that by black boxing individual units of data processing, the role of each process becomes very clear, and it is easy to build and maintain.

This is the same reason not only on the edge device, but also on the server side (cloud side), for persisting, analyzing, and visualizing the data that's collected by the edge device.

The biggest feature of Node-RED is that it connects the processing of Node.js in a sequential manner or in parallel with input/output data chunks in the form of messages. It can be said that this is very suitable for IoT data handling.

Again, as we discussed in the previous chapter, the standard architecture for IoT solutions is built on edge devices and cloud platforms. It sends the sensor data acquired by the edge device to the cloud, makes it persistent, and processes it for the desired processing chain.

This chapter will focus on that part of the cloud.

The edge device and the cloud don't actually connect yet. Assuming that the data has been passed to the cloud, let's make the data persistent in the database and visualize it.

We're going to use a dashboard node that is popular with all developers for Node-RED on IBM Cloud.

Before you use Node-RED on IBM Cloud, please install a new node; that is, **node-red-dashboard**.

Node-RED provides the **palette manager**, which is easy to install and is used to install extra nodes directly. This is very helpful when you're using lots of nodes. However, it might have issues due to the limited memory of the Node-RED application of an IBM Cloud Lite Account.

So, here, we need to get the **node-red-dashboard** node in order to edit the application's `package.json` file and redeploy the Node-RED application on IBM Cloud.

You can read about this node at `https://flows.nodered.org/node/node-red-dashboard`.

Follow these steps to make changes in the `package.json` file:

1. On the Node-RED **App details** page of IBM Cloud, click **source**. This will redirect you to a Git repository where you can edit the Node-RED application source code:

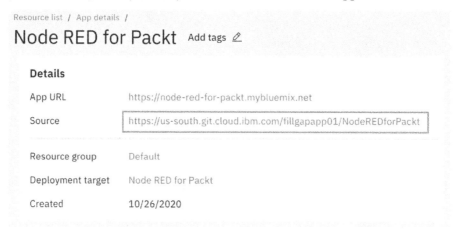

Figure 6.17 – Accessing your application source

2. Click on `package.json` on the file list. This file defines the module dependencies of your application:

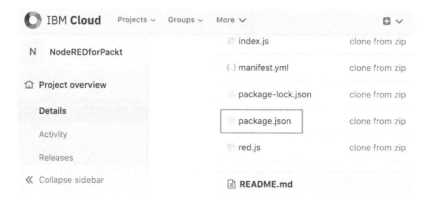

Figure 6.18 – Selecting package.json

3. Click the **Edit** button and add the following entry to the `dependencies` section:

```
"node-red-dashboard": "2.x",
```

4.   Add any commit message and click on the **Commit changes** button:

Figure 6.19 – Editing package.json and adding node-red-dashboard

After this, the Continuous Delivery Pipeline will automatically start to build and deploy the Node-RED application. You can check the status on the Delivery Pipeline at any time, just like you did while creating Node-RED Starter App:

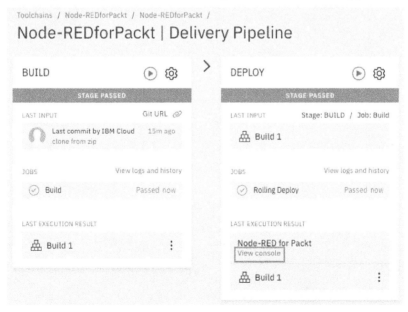

Figure 6.20 – Rebuilding and redeploying your application automatically

When you get **Deploy Stage** failed with the memory limit error for lite account, please stop your Node-RED service on your IBM Cloud dashboard and after that run the **Deploy Stage**. You can stop your Node-RED service by accessing your IBM Cloud dashboard and clicking on **Cloud Foundry apps** under **Resource summary**:

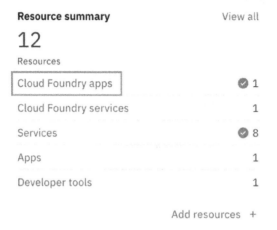

Figure 6.21 Selecting Cloud Foundry apps

After that, click on the **stop** option on the Node-RED record under the **Cloud Foundry apps**.

Figure 6.22 Clicking the Stop option

That's all. You can confirm that the dashboard node has been added by closing the **Palette management** screen and scrolling down the left-hand side of the flow editor, as shown in the following screenshot:

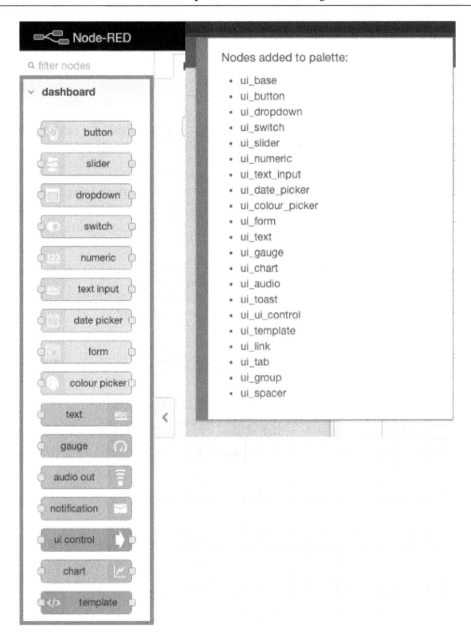

Figure 6.23 – Checking that the dashboard node has been installed

There's one more thing: we need to use a database, but IBM Cloud's version of Node-RED has a Cloudant database by default. We will use Cloudant for the case study in the next section.

Now, you can use Node-RED on IBM Cloud for IoT server-side situations.

IoT case study spot on the server side

Now, let's consider a server-side case study for IoT.

It does not depend on the case of each edge device. It primarily serves to process data and store it in a database for visualization.

In this chapter, we'll consider the use case of IoT; that is, assuming that the sensor data that's received using the sensor module is received on the server side, and the subsequent processing part.

The difference from the previous chapter is that in this server-side processing tutorial, the content of the data doesn't make much sense. The main purpose is to save the received data and visualize it as needed, so I would like to define the following two use cases.

Use case 1 – Storing data

The first case is to store data. Let's create an application (flow) that stores data you receive from devices. In this section, we don't use real data from devices; we just use the data generated by the inject node instead:

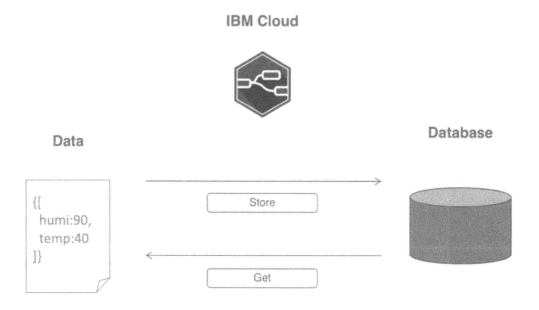

Figure 6.24 – Use case 1 overview

Now, let's look at the second use case.

Use case 2 – Temperature/humidity sensor

The second case is to show data as graphs or charts. Let's create an application (flow) that publishes data you received from devices, on the dashboard. We won't be using real data from any devices, just the data that's been generated by the inject node:

Figure 6.25 – Use case 2 overview

As we mentioned earlier, we will use Cloudant for the database for case 1 and the dashboard for the graph display for case 2. These have already been prepared.

Making a sample flow

Now, let's create these two server-side case flows on the Node-RED flow editor.

Please check again that the **Cloudant** node and the **Dashboard** node have already been installed on your flow editor. If you don't have them, please install these nodes by following the steps mentioned in the *What is the specific situation for using Node-RED on the cloud?* section of this chapter.

Now, you need to prepare a specific database for this tutorial on **Cloudant**. Follow these steps:

1. Access your IBM Cloud dashboard and click **View all** from the **Resource summary** area:

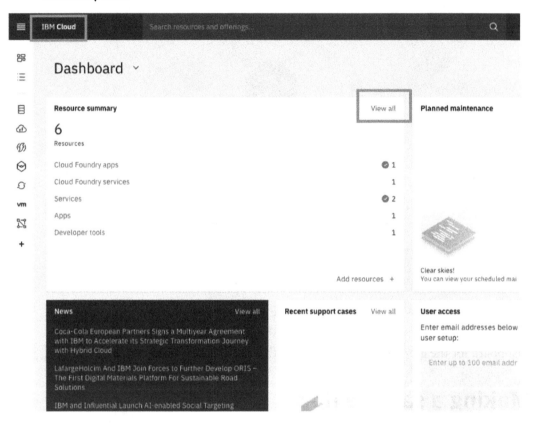

Figure 6.26 – IBM Cloud dashboard view

2. You will find the **Cloudant** service that you created using Node-RED. Please click the service's name:

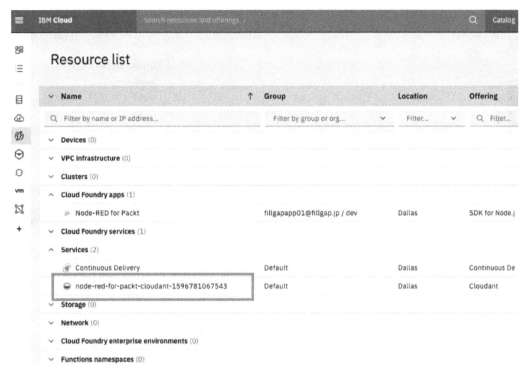

Figure 6.27 – Selecting the Cloudant service from the Resource list

3. Click the **Launch Dashboard** button at the top left of **IBM Cloud**:

Figure 6.28 – Launching the Cloudant dashboard

4. After launching the Cloudant dashboard, please click **Create Database** and enter a name for your database. You can name it whatever you want; here, we have used `packt_db`. After that, click the **Create** button:

Figure 6.29 – Creating a new database on Cloudant

Now that you have created the database for this tutorial, you can use it at any time!

Making a flow for use case 1 – storing data

With IoT, server-side processing starts from the point where it is received from the edge device. However, as we mentioned earlier, we will focus on storing the data in the database, so we will be using dummy data that will be generated by the **inject** node. The chunk of data that's received as a message is persisted in the Cloudant database on Node-RED.

We can make the flow by following these steps:

1. Place an **inject** node and a **cloudant out** node from the palette on the left-hand side of the flow editor by dragging and dropping them into the workspace:

Figure 6.30 – Placing the Inject node and cloudant out node

The **inject** node generates dummy data, while the **cloudant out** node stores the input value as-is in the Cloudant database.

2. After that, we will also create a flow to retrieve data from Cloudant, but first, let's just create the flow for saving data. Wire these nodes:

Figure 6.31 – Wiring these two nodes

3. Next, modify the settings of the **inject** node. Double-click the node to open the **Settings** panel.

4. Select **JSON** for the first parameter; that is, **msg.payload**, and click the right-hand side [...] button to open the JSON editor:

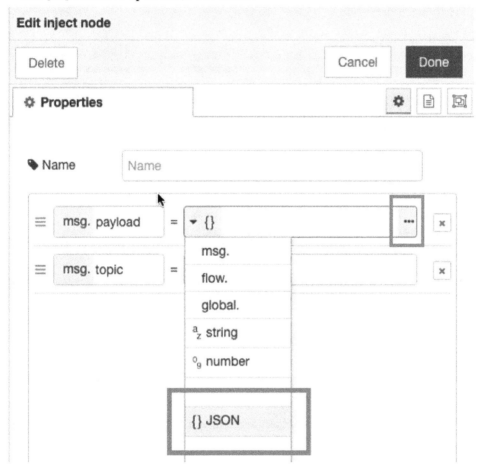

Figure 6.32 – JSON on the first parameter of the inject node

You can use both types of editor here; that is, the Text editor or the Visual editor. You can add any values to the JSON style, but here's what we have used for the JSON data:

```
{"temp":"29.18", "humi":"55.72"}
```

You can switch between the Text editor and the Visual editor using tabs. Please refer to the following image:

Figure 6.33 – Two types of JSON editor are available

There's no need to edit **msg.topic**.

5. After setting the **JSON** data, click the **Done** button in the top-right corner to close the **Settings** panel.

6. Then, edit the settings for the **cloudant out** node. This is simple: just enter `packt_db` as the database name. This name is the database you named on the Cloudant dashboard.

    The first parameter, **Service**, is set automatically; it is your Cloudant service on IBM Cloud. The third parameter, **Operation**, does not need to be changed from its default value.

7. After setting the database name, click the **Done** button in the top-right corner to close the **Settings** panel:

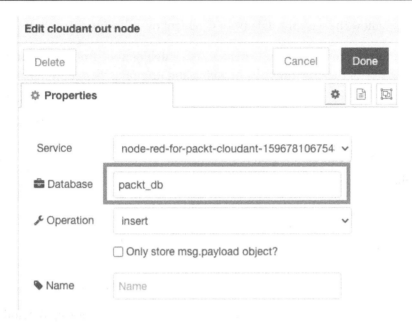

Figure 6.34 – Setting the database name on the cloudant out node

8. That's it! Don't forget to click the **Deploy** button.

9. Click the button on the **inject** node to run the flow. The data will be stored on the Cloudant database when the button has been clicked.

At this point, we can't check the data on Cloudant via the Node-RED flow editor; we can only check it on the Cloudant dashboard:

Figure 6.35 – Result on the Cloudant dashboard

Now, let's make a flow that gets the data from Cloudant by following these steps:

1. Place an **inject** node, a **cloudant in** node, and a **debug** node from the palette on the left-hand side of the flow editor by dragging and dropping them into the workspace from the previous flow.

   The **inject** node just executes this flow as a trigger, so there's no need to change the parameters in it. The **cloudant in** node gets the data from your Cloudant database. The **debug** node outputs a log on the debug tab.

2. Wire these nodes:

Figure 6.36 – Placing new three nodes and wiring them to get data

3. Next, modify the settings of the **cloudant in** node by double-clicking the node to open its **Settings** panel.

4. Just like the **cloudant out** node, enter `packt_db` as the database's name and select **all documents** for the third parameter' that is, **Search by**.

   The first parameter, **Service**, is set automatically; it is your Cloudant service on IBM Cloud.

5. After setting the database name and search target, click the **Done** button in the top-right corner to close the **Settings** panel:

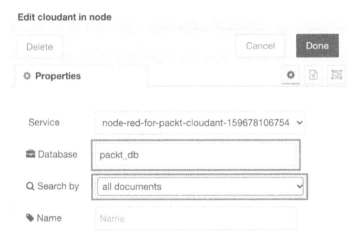

Figure 6.37 – Setting the database name and searching for a target on cloudant in the node

6. That's it! Don't forget to click the **Deploy** button.

7. Click the button on the **inject** node to run the flow. You will get the data from the Cloudant database when you do so.

You will see that the result was output to the **debug** window:

Figure 6.38 – Result of getting the data from Cloudant

Congratulations! With this, we have successfully created a basic flow (application) that stores sensor data on a database with Node-RED.

You can also download this flow definition file here: `https://github.com/PacktPublishing/-Practical-Node-RED-Programming/blob/master/Chapter06/cloudant-flows.json`

> **Important Note**
> This flow has no values for the Cloudant service name in cloudant in/out flows. Please check if your service name is set on that automatically once this flow definition has been imported.

You now understand how to handle data on Node-RED. We'll visualize that data in the next section.

Making a flow for use case 2 – visualizing data

The first use case was for storing sensor data in a database, while the second one was for visualizing sensor data on Node-RED. In IoT, after acquiring sensor data, we must visualize it in some form. The focus here is on retrieving and visualizing the data stored in use case 1. We will do this by following these steps:

1. Place an **inject** node, a **function** node, and a **chart** node from the palette on the left-hand side of the flow editor by dragging and dropping them into the workspace. Then, wire these nodes:

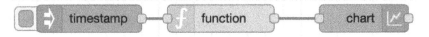

Figure 6.39 – Placing the nodes and wiring them to show data

The **Inject** node just executes this flow as a trigger, so there's no need to change the parameters in it. The **function** node generates numeric data to be shown on the Node-RED as a chart. Finally, the **chart** node makes it possible for the data to appear on the chart.

2. Code in the **function** node to generate numeric data that can be passed to the chart node.

3. Double-click the node to open the settings panel. Then, add the following code to the **function** node you placed:

```
// Set min and max for random number
var min = -10 ;
var max = 10 ;


// Generate random number and return it
msg.payload = Math.floor( Math.random() * (max + 1 - min)
) + min ;
return msg;
```

This is what it looks like:

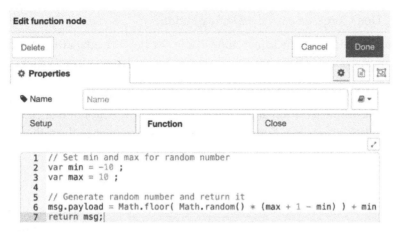

Figure 6.40 – Code for generating a random number

4. After coding this script, click the **Done** button in the top-right corner to close the **Settings** panel.

5. Then, edit the settings for the **chart** node. When the **Settings** panel opens, click the **pencil** button to the right of the **Group** parameter. The **dashboard group** settings screen will open. You can use the default name if you wish, but we named it `Packt Chart` here.

6. After entering a name, click the **Add** button in the top right to return to the **chart** node's settings panel; make sure the **Group** parameter is `Packt Chart`. Now, click the **Done** button at the top right:

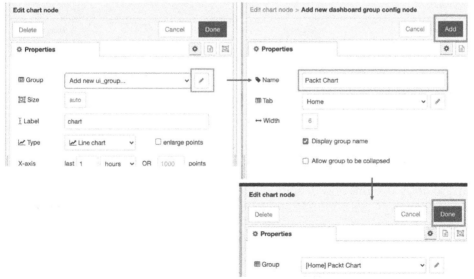

Figure 6.41 – Setting a parameter on the chart node

7. That's it! Don't forget to click the **Deploy** button.

8. Click the left button on the **inject** node to run the flow. The data that is generated by the **function** node will be sent to the **chart** node when the button is clicked.

    You can check the result on the dashboard window.

9. Click the **Dashboard** button at the top right of the flow editor and click the **Open window** button. These two buttons are icons, so please refer to the following screenshot to see which buttons you must click:

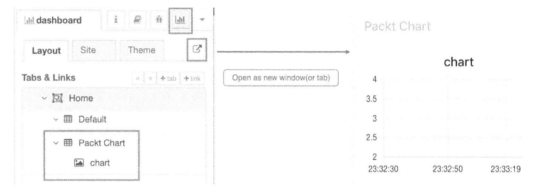

Figure 6.42 – Clicking the dashboard icon button and opening the window icon button

10. The line chart will be empty in the new window. Please click the switch of the **inject** node a few times. After that, you will see the line chart filled in with values:

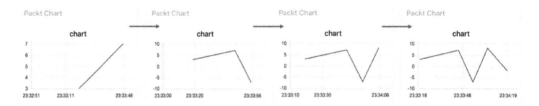

Figure 6.43 – Line chart with values

Congratulations! With this, we have successfully created a basic flow (application) that shows sensor data as a chart with Node-RED.

You can also download this flow definition file here: `https://github.com/ PacktPublishing/-Practical-Node-RED-Programming/blob/master/ Chapter06/dashboard-flows.json`.

Summary

In this chapter, you learned how to create a server-side sample flow (application) by following a real IoT use case. These were simple tutorials, but I am sure it will be beneficial for you so that you understand how to make flows for IoT server-side applications.

The flow steps we created here will help you create different flows for other server-side applications in the future.

In the next chapter, we will use the same IoT use case we used in this chapter, but we will create a practical sample flow (application) that will call a web API.

7
Calling a Web API from Node-RED

In this chapter, let's call a web API from Node-RED. Basically, in Node-RED, processing is performed as per the created flow, but it is JSON data that connects processing. In that sense, it is very compatible with web APIs.

Let's get started with the following four topics:

- Learning about the RESTful API
- Learning about the input/output parameters of a node
- How to call the web API on a node
- How to use the IBM Watson API

By the end of this chapter, you will have mastered how to call any type of web API from Node-RED.

Technical requirements

To progress through this chapter, you will need the following:

- Node-RED (v1.1.0 or above)

The code used in this chapter can be found in `Chapter07` folder at `https://github.com/PacktPublishing/-Practical-Node-RED-Programming`.

Learning about the RESTful API

Many of you reading this book may already be familiar with web APIs. However, let's review the RESTful API in order to call a web API with Node-RED.

REST stands for **Representational State Transfer**. RESTful API basically refers to the invocation interface in HTTP of a web system that is implemented according to "REST principles." So, in a broad sense, it's safe to say that the REST API and RESTful API are the same things. So, what exactly is the RESTful API? We will learn the outline and principles of the RESTful API, and the advantages and disadvantages of using the RESTful API, in this section.

REST was proposed by Roy Fielding, one of the HTTP protocol creators, around the year 2000, and is a set (or way of thinking) of design principles suitable for linking multiple software when building a distributed application. In addition, the RESTful API is an API designed according to the following four REST principles:

- **Addressability**: It has the property of being able to directly point to a resource through a URI. All information should be represented by a unique URI so that you can see at a glance the API version, whether to acquire data, update, and so on.

- **Statelessness**: All HTTP requests must be completely separated. State management such as sessions should not be performed.

- **Connectivity**: This refers to the ability to include a "link to other information" in one piece of information. By including a link, you can "connect to other information."

- **Unified interface**: Use HTTP methods for all operations such as information acquisition, creation, update, and deletion. The HTTP methods, in this case, are acquisition ("GET"), creation ("POST"), update ("PUT"), and deletion ("DELETE").

These are the four principles. As you can see from these four principles, a major feature of REST is that it makes more effective use of HTTP technology and has a high affinity with web technology. Therefore, it is currently used for developing various web services and web applications.

With the recent widespread use of smartphones, it is becoming more obvious that business systems can be used not only on PCs but also on mobiles. In addition, not just one system but a system that can be linked with multiple systems and various web services will not be selected by users. RESTful APIs are receiving a great deal of attention as an indispensable tool for solving these problems.

As the following figure shows, a web API can be called from anywhere via the internet:

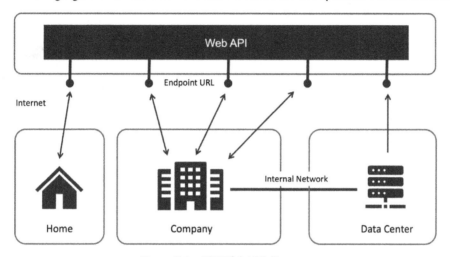

Figure 7.1 – RESTful API diagram

Now, let's recall what Node-RED is. Its workflow tool-like style is like a standalone tool, but Node-RED is certainly one of web applications too. In other words, it's an application that works very well with the RESTful API described here.

Next, let's cover again what kinds of parameters Node-RED nodes have.

Learning about the input/output parameters of a node

Of the many nodes that Node-RED has, not many are suitable for calling web APIs (REST APIs). A typical node used when calling the web API is the http request node.

To call an external API on Node-RED, simply set the endpoint URL of the API to the URL property of the http request node.

For example, when it is necessary to set a parameter in the endpoint URL when calling an API, it is possible to set the output value of the previous node connected. The method is very easy. Instead of a literal string, you can just set the {{payload}} variable in the value setting part of the parameter.

In {{payload}}, the character string inherited from the previous processing node is entered.

Take the following example (note that this URL does not exist): http://api-test. packt.com/foo?username={{payload}}&format=json:

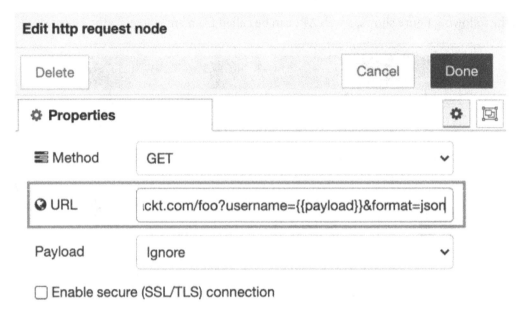

Figure 7.2 – Setting the API endpoint URL with {{payload}} as a parameter

The process of the http request node cannot be executed by the http request node alone. Before the http request node, it is necessary to connect the trigger process, such as the inject node. At that time, if there is a parameter you want to pass to the API call, that is, the http request node, please set it in msg.payload.

If the API you want to call in the http request node is POST, the JSON data to be included in the request will be satisfied as a request parameter by creating it in the preprocessing node, storing it in msg.payload as it is, and connecting it to the http request node.

By using the http request node like this, API cooperation can be easily realized. API calls are important for linking multiple services on Node-RED. For example, the **function** node of Node-RED is basically processed by JavaScript, but by making a program developed in other development languages, such as Java, into an API, it can be used by calling from Node-RED.

How to call the web API on a node

So far, we've learned what a RESTful API is and which node is appropriate for an API call.

In this part, let's create a flow that actually calls the API from Node-RED and learn how to call the API and how to handle the result value from the API.

There are a few things to think about first, such as which API to call. Fortunately, various APIs are published on the internet.

This time, I would like to use the OpenWeatherMap API. In OpenWeatherMap, for example, the following APIs for data acquisition are prepared:

- Current weather data
- Hourly forecast 4 days
- Daily forecast 16 days
- Climatic forecast 30 days
- Weather alerts
- And more...

For more information, please see the official website of OpenWeatherMap: `https://openweathermap.org/`.

Okay, let's prepare to use the OpenWeatherMap API.

Creating an account

To use the OpenWeatherMap API, we need to create an account. Please access the following URL: `https://openweathermap.org/`.

If you already have an account, please log in without taking the following steps.

For those who are using it for the first time, please click the **Sign In** button, and then click the **Create an Account** link. It is easy to register. Just follow the guidance and confirm the email sent to you by OpenWeatherMap after registration. This is what the creating an account page looks like:

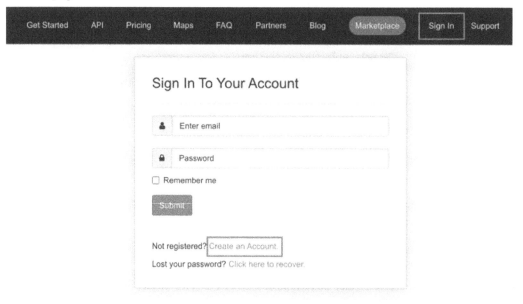

Figure 7.3 – Creating an OpenWeatherMap account

Next, let's create an API key.

Creating an API key

When you log in to OpenWeatherMap, you can see the **API keys** tab, so please click it. You already have a default API key but please create a specific API key for this tutorial. Enter any key string and click the **Generate** button.

Please note that the API keys shown in this book are created by me as a sample and cannot be used. Be sure to create a new API key in your account:

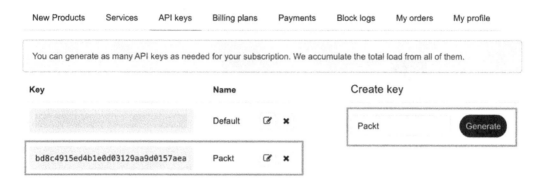

Figure 7.4 – Generating API key

> **Important note**
>
> After creating the API key, the key will not be activated for 10 minutes to a couple of hours. If a web response error such as 401 is returned even when you access the API endpoint URL described in the next section, the specified API key may not have been activated, so please wait and try again.

Checking the API endpoint URL

To check your API endpoint URL, follow these steps:

1. Click the **API** button on the menu bar. You can see some APIs there.

2. In this tutorial, we will use **Current Weather Data**, so please click the **API doc** button under **Current Weather Data**:

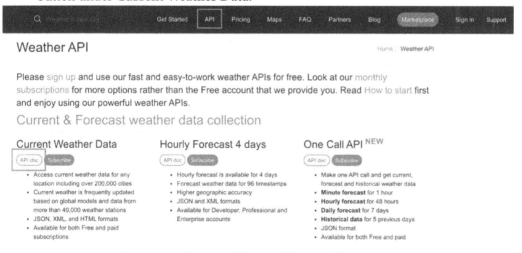

Figure 7.5 – Opening API doc of Current Weather Data

3. This API has some types of parameters such as **By city**, **By city ID**, **By zip code**, and so on. Please select **By city name** with the parameter city name and API key.

    API doc, **city name**, **state code**, and **country code** are from ISO 3166. The URLs under the **API call** area are endpoint URLs for using this API. Please copy this URL to the clipboard:

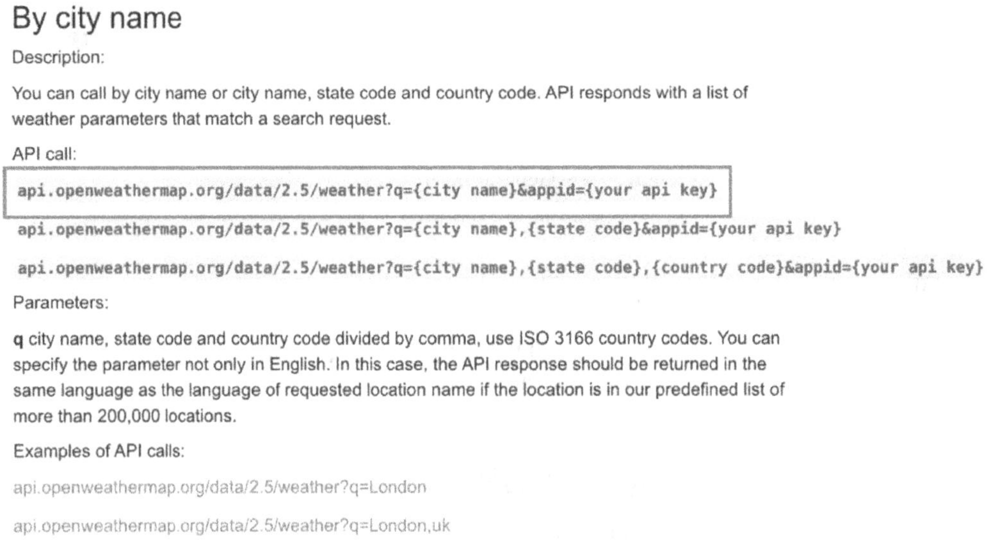

By city name

Description:

You can call by city name or city name, state code and country code. API responds with a list of weather parameters that match a search request.

API call:

```
api.openweathermap.org/data/2.5/weather?q={city name}&appid={your api key}
```

```
api.openweathermap.org/data/2.5/weather?q={city name},{state code}&appid={your api key}
```

```
api.openweathermap.org/data/2.5/weather?q={city name},{state code},{country code}&appid={your api key}
```

Parameters:

q city name, state code and country code divided by comma, use ISO 3166 country codes. You can specify the parameter not only in English. In this case, the API response should be returned in the same language as the language of requested location name if the location is in our predefined list of more than 200,000 locations.

Examples of API calls:

api.openweathermap.org/data/2.5/weather?q=London

api.openweathermap.org/data/2.5/weather?q=London,uk

Figure 7.6 – API endpoint URL with parameters

Next, let's see whether we can run this API or not.

Checking that the API can run

Let's try to use this API. You just have to open your browser, paste the URL, and replace the city name and API key with yours. You can choose any city name, but the API key is the specific one you created in the previous section:

{"coord":{"lon":139.69,"lat":35.69},"weather":[{"id":802,"main":"Clouds","description":"scattered clouds","icon":"03d"}],"base":"stations","main":{"temp":304.7,"feels_like":310.57,"temp_min":303.15,"temp_max":306.48,"pressure":1013,"humidity":79},"visibility":10000,"wind":{"speed":3.1,"deg":150},"clouds":{"all":40},"dt":1598573164,"sys":{"type":1,"id":8077,"country":"JP","sunrise":1598559021,"sunset":1598606104},"timezone":32400,"id":1850144,"name":"Tokyo","cod":200}

Figure 7.7 – Calling the API and getting the result

I have now confirmed that this API works correctly. Now let's call this API from Node-RED and use it.

Creating the flow calling the API

Now let's create a flow that calls the OpenWeatherMap API on Node-RED. Start Node-RED in your environment. You can use either standalone Node-RED or Node-RED on IBM Cloud:

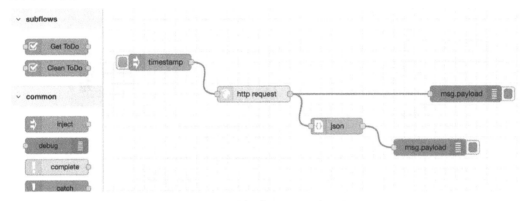

Figure 7.8 – The flow to use the API

For this, the flow is very simple and easy to make. Please follow these steps to make the flow:

1. Place one **inject** node and two **debug** nodes on the palette. These nodes can be used as default. No change in the settings is required here.

2. Place the **http request** node on the palette, then open the settings panel of the **http request** node and set the API endpoint URL with your parameters (city name and API key) in the **URL** textbox of the settings panel, as shown in the following figure:

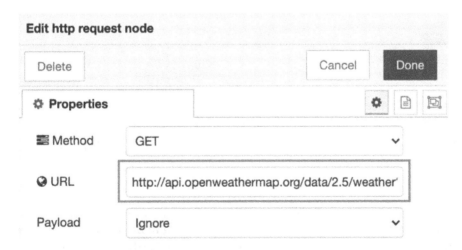

Figure 7.9 – Setting the API endpoint URL with your parameters

3. Place a **json** node on the palette. This node can be used with the defaults. No changes in the settings are required here. But, just in case, let's make sure that the **Action** property of the **json** node is set to **Convert between JSON String & Object**. This is an option that will convert the JSON data passed as the input parameter into a JavaScript object:

Figure 7.10 – Checking the Action property

4. Wire all nodes as shown in the following figure:

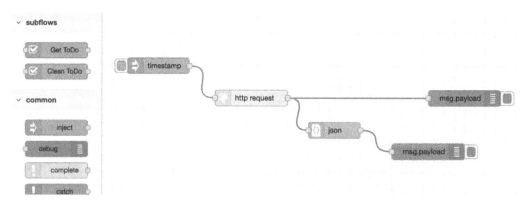

Figure 7.11 – Wiring all nodes

Please wire the **timestamp** node and the **http request** node. The **http request** node output is wired to the **json** node and the **debug** node. Lastly, please wire the **json** node output to another **debug** node.

5. After changing the settings and wiring all the nodes, you need to deploy and click the switch of the **inject** node. You can now see the data on the **debug** window in the right-side panel:

Figure 7.12 – Result data (JSON) on the debug window

You can also see the result data as a JSON object on the same **debug** window as in the following screenshot:

Figure 7.13 – Result data (object) on the debug window

Congratulations! You have succeeded in making a sample flow by calling the OpenWeatherMap API. If you didn't succeed in making this flow completely, you can also download this flow definition file here: `https://github.com/` `PacktPublishing/-Practical-Node-RED-Programming/blob/master/` `Chapter07/open-weather-flows.json`.

In the next section, we will learn about the convenience of using the IBM Watson API with Node-RED on IBM Cloud.

How to use the IBM Watson API

In the previous section, you learned how to call the API and handle the resulting values from the API.

In this section too, we will create a flow that actually calls the API from Node-RED, but we will learn how to call the Watson API provided by IBM. We will also create a flow that actually calls the API from Node-RED, but we will learn how to call the Watson API provided by IBM.

Why Watson? Watson is a brand of artificial intelligence services and APIs provided by IBM.

All Watson APIs can be used from IBM Cloud. So, by running Node-RED on IBM Cloud, you can effectively use Watson's services. This has advantages such as when calling the Watson API from Node-RED, implementation of authentication can be omitted.

Watson can be called from environments other than IBM Cloud, so it can be called directly from a Raspberry Pi or can be used from either cloud platforms such as AWS and Azure or on-premises environments. See the following figure, showing what a Watson API looks like:

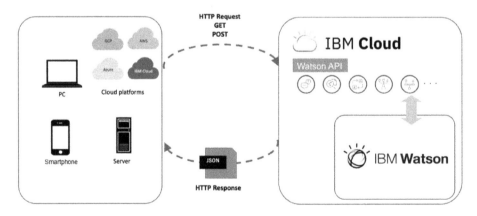

Figure 7.14 – Watson API diagram

For more information, see the IBM Watson official website: `https://www.ibm.com/watson`.

Okay, let's see how easy it is to use the Watson API on Node-RED on IBM Cloud.

Logging in to IBM Cloud

If you've followed the steps from the first chapter, you should already have an IBM Cloud account. Just log in to IBM Cloud (`https://cloud.ibm.com`).

If you do not have an IBM Cloud account, create one from the following URL and log in to IBM Cloud. See *Chapter 6, Implementing Node-RED in the Cloud*, for detailed instructions: `http://ibm.biz/packt-nodered`.

Starting Node-RED on IBM Cloud

In the previous section, we created a sample flow using standalone Node-RED or Node-RED on IBM Cloud. Of course, you can use the standalone version of Node-RED to call the Watson API, but some benefits will be lost. So, we will work with Node-RED on IBM Cloud in this part.

As in the previous step, if you have not used Node-RED on IBM Cloud yet, please return to *Chapter 6, Implementing Node-RED in the Cloud*, and run through it to activate Node-RED on IBM Cloud before moving on to the next step.

Creating the Watson API

Next, create Watson's service on IBM Cloud. Strictly speaking, this means creating an instance as a service so that you can call the Watson API service provided on IBM Cloud as your own API.

Watson has several APIs, such as voice recognition, image recognition, natural language analysis, sentiment analysis, and so on. This time, I would like to use the sentiment analysis API.

Follow these steps to create a Watson Tone Analyzer API service:

1. Search for Watson from the catalog. On the dashboard, please click the **Catalog** menu item and search for tone analyzer, and then select the **Tone Analyzer** panel:

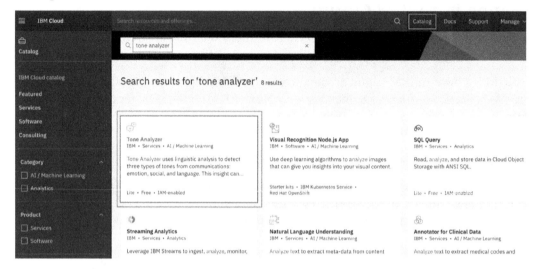

Figure 7.15 – Searching Watson services

2. Please refer to the following list and *Figure 7.16* to fill in each property:

   a. **Region**: **Dallas** (you can select any region, but Dallas is recommended)

   b. **Pricing plan**: **Lite** (free pricing)

   c. **Service name**: Default (you can modify this to any name you want to use)

   d. **Resource group**: **Default** (you can't select anything else for a Lite account)

   e. **Tags**: N/A

3. After entering/selecting all the properties, click the **Create** button:

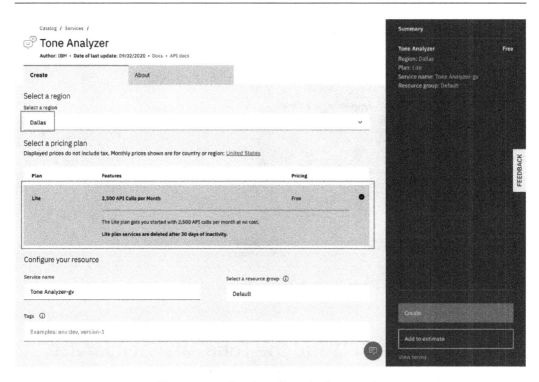

Figure 7.16 – Creating a Tone Analyzer service

4. You can see the status as **Active** on the **Tone Analyzer** instance dashboard when it is created and activated. Please check the API key and URL. API keys and URLs are used when the API is called from any application. However, these are not used in this tutorial because Node-RED on IBM Cloud can call the Watson API without authentication coding.

You can check the API key and URL from the **Manage** menu on this screen:

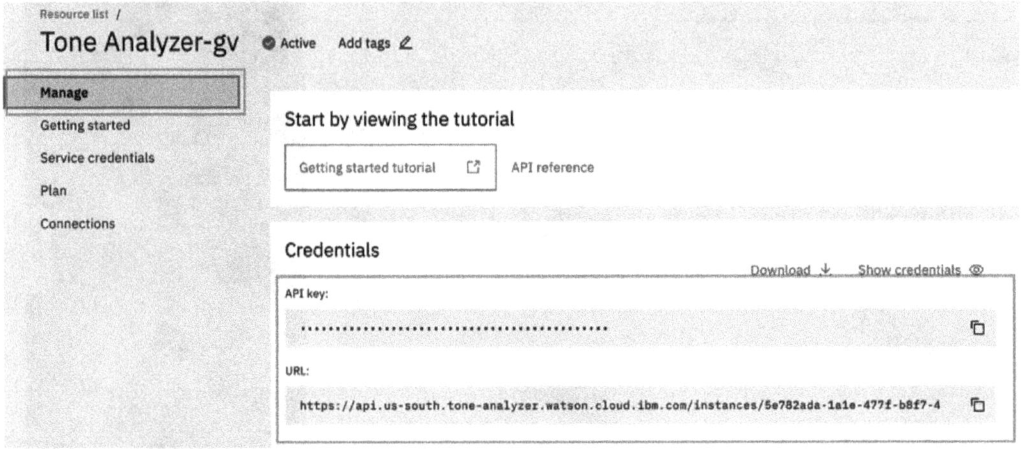

Figure 7.17 – Checking your credentials

In the next section, we will connect Node-RED and the Tone Analyzer service.

Connecting Node-RED and the Tone Analyzer service

As you already know, Node-RED can call the Watson API without coding for authentication. We need to connect Node-RED and the Watson API instance before using Node-RED with the Watson API. In the last step, we created the **Tone Analyzer** API instance, so let's connect these two instances as follows:

1. Click the **IBM Cloud** logo button at the top left to move to the main dashboard.

2. Click the **View all** button on the **Resource summary** panel.

3. Click the Node-RED instance (application) name in the **Cloud Foundry apps** area:

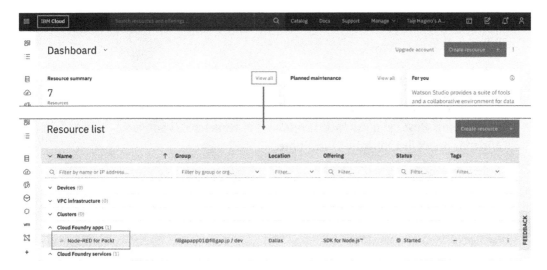

Figure 7.18 – Selecting the Node-RED service you created

4. Click the **Connections** menu and then the **Create connection** button:

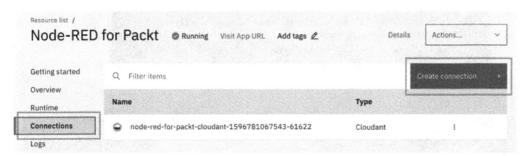

Figure 7.19 – Creating a connection for Node-RED and Watson

5. Check the **Tone Analyzer** service and click the **Next** button:

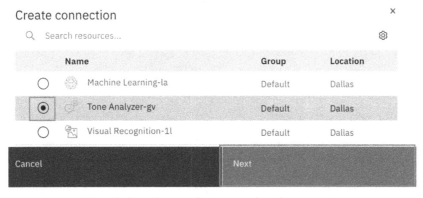

Figure 7.20 – Clicking the Next button to select the connecting service

6. No modification is needed for the access role and service ID. Click the **Connect** button:

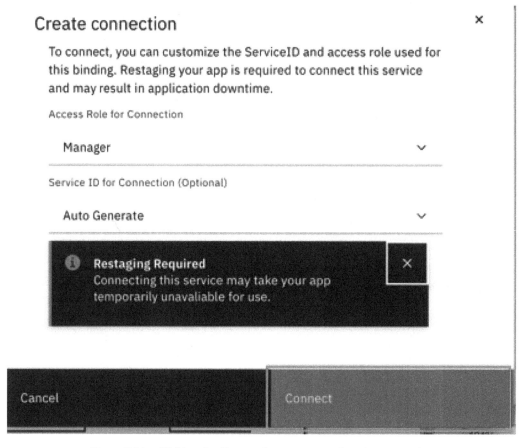

Figure 7.21 – Clicking the Connect button to complete the connection

7. We need to restage Node-RED to activate the connection. Click the **Restage** button:

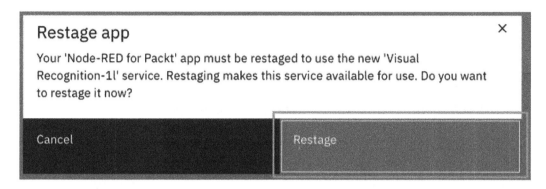

Figure 7.22 – Clicking the Restage button to start restaging the Node-RED service

8. Please wait until the restaging of your Node-RED instance is completed. Once completed, you will get a successful connection with the **Running** status. After that, please open the Node-RED flow editor via the **Visit App URL** link:

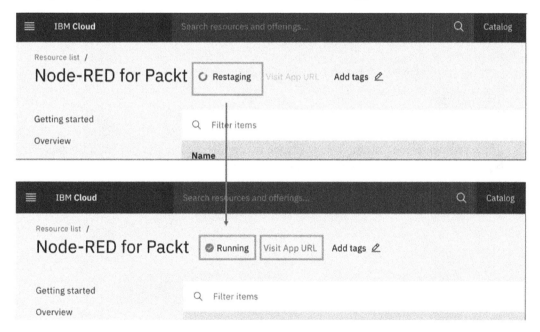

Figure 7.23 – Checking the status of the Node.js runtime for the Node-RED service

You have succeeded in preparing the Node-RED and Watson API flow. Next, let's create the flow by calling the Tone Analyzer API.

Creating the flow by calling the Tone Analyzer API

Now, let's create a flow that calls the Watson Tone Analyzer API on Node-RED. You already started Node-RED on IBM Cloud. Either standalone Node-RED or Node-RED on IBM Cloud can be used.

To proceed with this tutorial, you need to install the following two nodes:

- **node-red-node-twitter**: This is a node that acquires and posts tweets to Twitter:

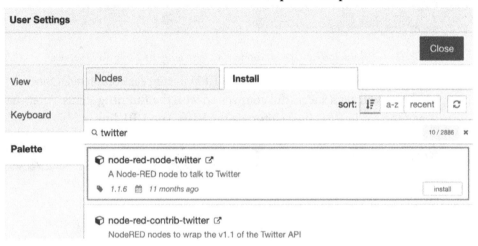

Figure 7.24 – Installing node-red-node-twitter

- **node-red-node-sentiment**: This is a node that adds a sentiment object in the passed `msg.payload`. It is used when passing parameters to the Watson Tone Analyzer API:

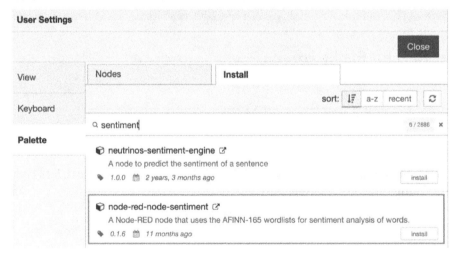

Figure 7.25 – Installing node-red-node-sentiment

Search for these nodes in the palette and install them to your Node-RED flow editor. After that, make a flow as shown in the following figure:

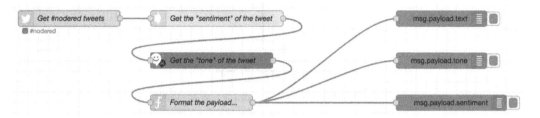

Figure 7.26 – The flow to use the Tone Analyzer API

In this flow, the function node processes the text, tone, and sentiment included in the result value obtained from Twitter so that they are output as separate debugs. This is to make the results easier to see.

This flow is a little more complicated than the flow you created in the previous step. Please follow these steps to make the flow:

1. Make a Twitter ID (Twitter account) and create an application on your Twitter Developer account to authenticate for accessing tweets via your Twitter account.

2. Access **Overview** under **Projects & Apps** on Twitter Developer, and then click the **Create an app** button:

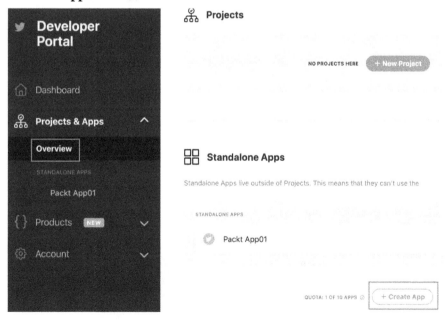

Figure 7.27 – Creating an app on Twitter Developer

3. Set the **App name** with any strings, and click the **Complete** button.

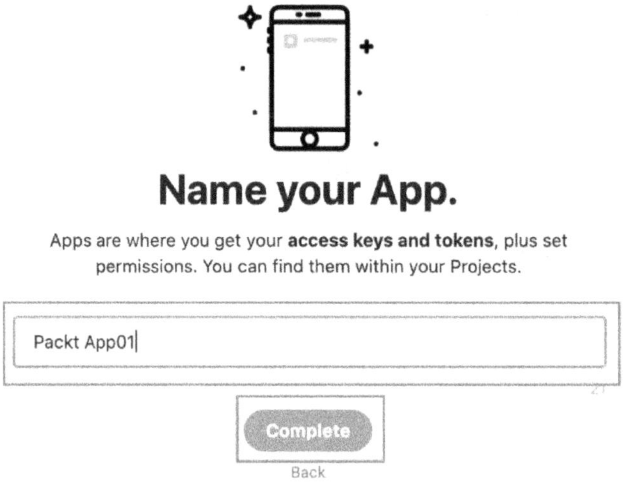

Name your App.

Apps are where you get your **access keys and tokens**, plus set
permissions. You can find them within your Projects.

Packt App01

Complete

Back

Figure 7.28 – Setting a name of your app

4. After that, please check the **Access token & access token secret** area.

    You will see the tokens. Please note and save your access token and access token
    secret. These will be used for the setting of the **twitter in** node too:

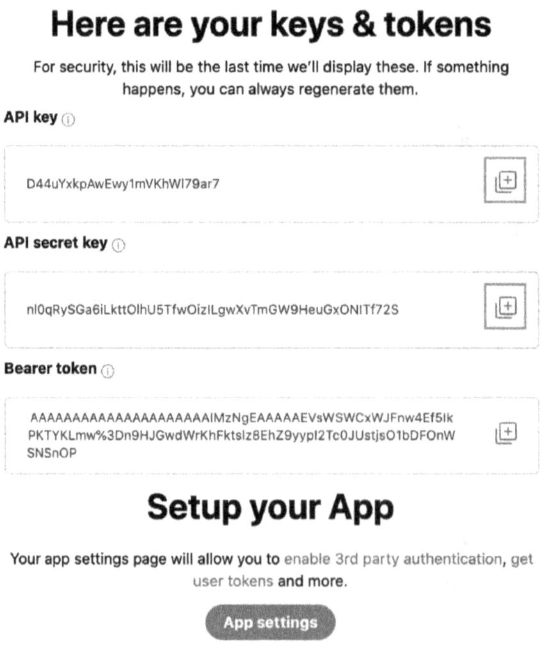

Here are your keys & tokens

For security, this will be the last time we'll display these. If something
happens, you can always regenerate them.

API key ⓘ

D44uYxkpAwEwy1mVKhWl79ar7

API secret key ⓘ

nI0qRySGa6iLkttOlhU5TfwOizlLgwXvTmGW9HeuGxONITf72S

Bearer token ⓘ

AAAAAAAAAAAAAAAAAAAAAIMzNgEAAAAAEVsWSWCxWJFnw4Ef5lk
PKTYKLmw%3Dn9HJGwdWrKhFktslz8EhZ9yypl2Tc0JUstjsO1bDFOnW
SNSnOP

Setup your App

Your app settings page will allow you to enable 3rd party authentication, get
user tokens and more.

App settings

Figure 7.29 – Note your token and token secret

5. Place the **twitter in** node on your workspace, and double-click it to open the settings window:

Figure 7.30 – Placing the twitter in node

6. Click the edit (pencil icon) button on the settings window to edit your Twitter information:

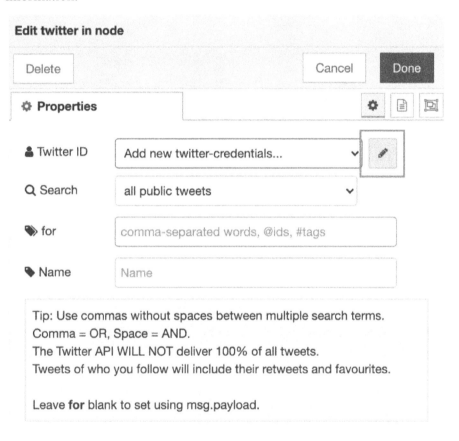

Figure 7.31 – Editing the Twitter properties

7. Set your Twitter ID, API key, and token.

    The values for **API key**, **API key secret**, **Access token**, and **Access token secret** should be taken from your text editor from *step 8*.

8. After setting these settings, please click the **Add** button to return to the main settings window of the **twitter in** node:

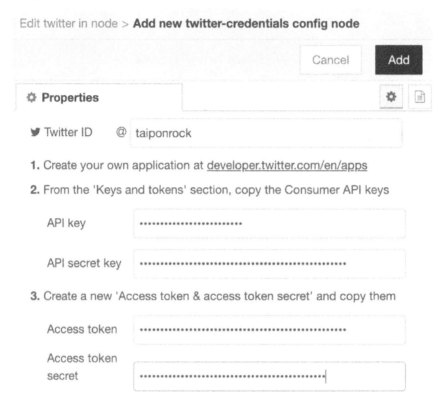

Figure 7.32 – Configuring your Twitter information

9. Select **all public tweets** for **Search**, and set **for** to #nodered as the hashtag. You can set any name for **Name**.

10. Finally, click the **Done** button to finish adding these settings and close the window:

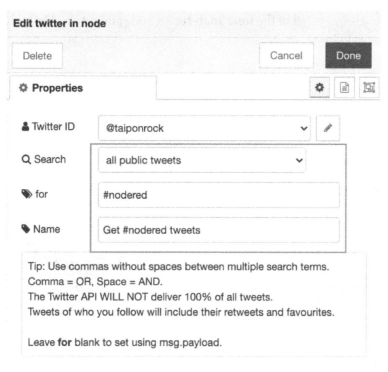

Figure 7.33 – Finalizing the settings of the twitter in node

11. Place the **sentiment** node on your workspace. It will be wired after the **twitter in** node.

For this node, no properties are needed to be set or changed:

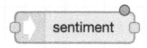

Figure 7.34 – Placing the sentiment node

12. Place the **tone analyzer v3** node after the **sentiment** node sequentially on your workspace:

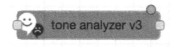

Figure 7.35 – Placing the tone analyzer v3 node

13. Open the settings panel of the **tone analyzer v3** node and set the **Method** and **URL** properties as follows:

a. **Name**: Any string you want to name

b. **Method: General Tone**

c. **version_date: Multiple Tones**

d. **Tones: All**

e. **Sentences: True**

f. **Content type: Text**:

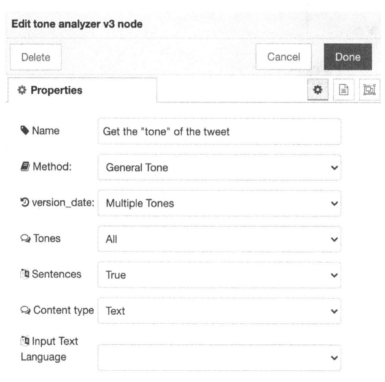

Figure 7.36 – Configuring the tone analyzer v3 node properties

14. Place the **function** node after the **tone analyzer v3** node sequentially on your workspace:

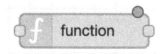

Figure 7.37 – Placing the function node

15. Open the settings panel of the **function** node and code JavaScript with the following
source code:

```
msg.payload = {
    "text" : msg.payload,
    "tone" : msg.response,
    "sentiment" : msg.sentiment
};
return msg;
```

Please refer to the following screenshot for the coding for the **function** node:

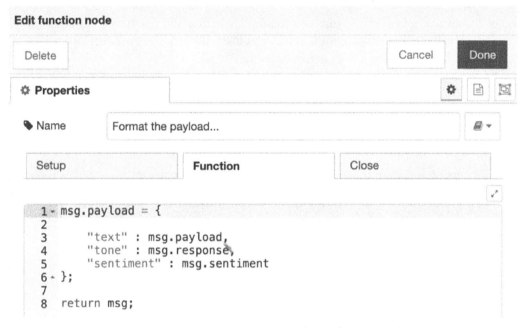

Figure 7.38 – JavaScript source code for the function node

You can get the code here: `https://github.com/PacktPublishing/-`
`Practical-Node-RED-Programming/blob/master/Chapter07/`
`format-payload.js`.

16. Finally, put three **debug** nodes in parallel after this **function** node. Each **debug** node's settings are as follows:

- `msg.payload.text`: For the **debug** tab

- `msg.payload.tone`: For the **debug** tab

- `msg.payload.sentiment`: For the **debug** tab

See *Figure 7.26* for the wiring instructions. We have finished making configurations for the nodes of the flow.

Testing the flow

The flow is now complete. When you click the **deploy** button and the `twitter in` node connects to Twitter using your account, it will automatically retrieve the tweets that meet your criteria and process the subsequent flow.

This is done automatically, so you don't have to take any special action.

Here, it is set to get all tweets that have #nodered as a hashtag. If you don't get many tweets, it means that a tweet that contains the specified hashtag has not been created, so please change the hashtag set in the `twitter in` node and try again.

All the processing results of this flow will be displayed in the **debug** tab.

It is `msg.payload.text` that extracts the tweet body from the acquired tweets and displays it:

Figure 7.39 – Result of getting the tweet body

It is msg.payload.tone that extracts and displays emotions detected in the acquired tweets:

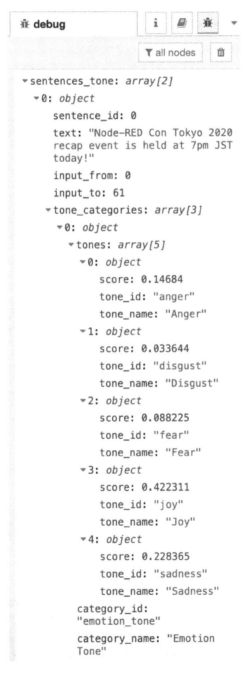

Figure 7.40 – Result of tone analysis from the tweet

It is `msg.payload.sentiment` that judges whether the sentiment is positive or negative in the acquired tweets:

Figure 7.41 – Result of the sentiment of a tweet

Congratulations! You have succeeded in making a sample flow by calling the Watson API. If you didn't succeed in making this flow completely, you can also download this flow definition file here: `https://github.com/PacktPublishing/-Practical-Node-RED-Programming/blob/master/Chapter07/get-sentiment-twitter-flows.json`.

Summary

In this chapter, we learned how to create a sample flow (application) that calls two types of web APIs. We are gradually getting used to creating complicated flows. Use cases for calling web APIs are frequently found in Node-RED. The flow creation methods we learned about here will help us to create more complex flows in the future.

In the next chapter, let's learn about a project feature that can be integrated with repositories such as GitHub, which is a function added from Node-RED version 1.0.

8
Using the Project Feature with Git

What you will learn in this chapter is a very useful **Project** feature. The project feature of Node-RED is a kind of version management tool with Git on a Node-RED flow editor. This is actually disabled by default. Enabling this allows you to manage your flows in a new way. I believe many developers are familiar with Git services such as GitHub and GitLab. The project feature of Node-RED uses Git and GitHub for version control, so I think it's very easy to understand.

Here are the topics that we will be covering in this chapter:

- Enabling the project feature
- Using the Git repository
- Connecting a remote repository

By the end of this chapter, you will be able to understand how to use the project feature, how to connect your own Git repository to your Node-RED flow editor, and how to manage flows as projects with version control tool Git.

By the end of this chapter, you will have mastered how to use the project feature and make your applications with it. You can use it in any hosted Git service such as GitHub or GitLab.

Technical requirements

To progress in this chapter, you will require the following:

- A GitHub account, which you can create via the official website: `https://github.com/`.

- A Git client tool, which you need to install via the official website: `https://git-scm.com/downloads`.

Enabling the project feature

For example, in situations where you want to manage your own flow while sharing it with others, or you want to update the flow created by others, it is difficult to develop when a team uses only the Node-RED flow editor.

The project function of Node-RED is a method/function for managing the files that are relevant with each flow you make. It covers all the files needed to create applications with Node-RED shareable.

These are supported by the Git repository. That is, all files are versioned. This allows developers to collaborate with other users.

On Node-RED version 1.x, the project feature is disabled by default, so it must be enabled in the `config` file named `settings.js`.

Important note

When creating a project in the local environment of Node-RED, the flow created so far may be overwritten with a blank sheet. You can download the JSON files of the flow configurations for all the flows created in this document via the internet, but if the flow you created yourself exists in Node-RED in the local environment, it is recommended to export the flow configuration file.

All of the flow definitions and JSON files that we created in this book are available to download here: `https://github.com/PacktPublishing/-Practical-Node-RED-Programming`.

Now let's try the project function. We will use a standalone version of Node-RED on a local environment such as macOS or Windows. In order to use the project feature, we first need to enable it. Let's enable it by following these steps:

1. It is necessary to rewrite the `settings.js` file to enable/disable the project function. Look for this file first. The `settings.js` file can be found in the Node-RED user directory where all of the user configurations are stored.

   By default, on a Mac, this file is available under the following path:

   `/Users/<User Name>/.node-red/settings.js`.

   By default, on Windows, this file is available under the following path:

   `C:\Users\<User Name>\.node-red\settings.js`

2. Edit the `settings.js` file. It is OK to open `settings.js` with any text editors. I have used `vi` here. Open `settings.js` with the following command:

   ```
   $ vi /Users/<User Name>/.node-red/settings.js
   ```

   > **Important note**
   > Please replace the command with the one specific to your environment.

3. Edit your `settings.js` file and set the **projects.enabled** element to `true` in the `editorTheme` block within the `module.exports` block in order for the project feature to be enabled:

   ```
   module.exports = {
       uiPort: process.env.PORT || 1880,

       ...

       editorTheme: {
           projects: {
               enabled: true
           }
       },

       ...

   }
   ```

4. Save and close the `settings.js` file.

5. Restart Node-RED to enable the settings we modified by running the following command:

```
$ node-red
```

We have now successfully enabled the project feature of Node-RED.

To use this feature, you need to have access to Git and ssh-keygen command-line tools. Node-RED checks them at startup and notifies you if any tools are missing.

If the settings are completed without any problems and you have restarted Node-RED, the project feature will be available. Next, let's set up the Git repository to use.

Using the Git repository

We enabled the project feature in the previous section. Reopen the flow editor and you will be prompted to create your first project using the contents of the flow you created at that time. This will be the welcome screen:

Hello! We have introduced 'projects' to Node-RED.

This is a new way for you to manage your flow files and includes version control of your flows.

To get started you can create your first project using your current flow files in a few easy steps.

If you are not sure, you can skip this for now. You will still be able to create your first project from the 'Projects' menu option at any time.

Figure 8.1 – Welcome screen

We need to set up a version control client such as Git. As already explained, the project function of Node-RED uses Git as a version control tool. As with regular Git, you can manage file changes on a project-by-project basis and synchronize with remote repositories as required.

Git keeps track of who made the change. It works with your username and email address. The username does not have to be your real name; you can use any name you like.

If your local device already has a Git client configured, Node-RED will look up those settings.

First, perform version control in your local environment. It takes advantage of the features of the Git client installed in your local environment. If you do not have Git installed, please install it in advance.

Now, follow these next steps to create a project on your Node-RED flow editor:

1. First, let's create a project. This is very easy. Enter a project name and a description in the project creation window.

2. Name the flow file. By default, it is already named `flow.json`.

    In other words, Node-RED automatically migrates the flow currently configured on the flow editor to a new project as it is. It is OK to keep the default name. Of course, you may choose to rename it here if you so wish.

    If you publish your project on a public site such as GitHub, it's a good idea to encrypt your credentials file.

    If you choose to encrypt, you must create a key to use for encryption. This key is not included in the project, so if you share the project with someone, you will need to provide the credential file decryption key separately to the user who cloned the project.

3. After adding the required information, click the **Create Project** button:

Figure 8.2 – Projects screen

Congratulations! You have created your first project.

4. Next, check the project history. We can use the version control feature on our Node-RED flow editor. You can access the project history panel by clicking the **project history** button in the top-right corner:

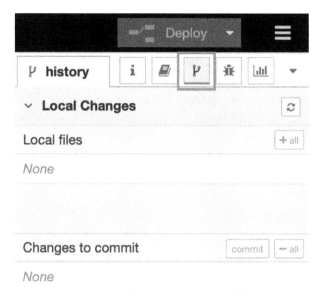

Figure 8.3 – Project history panel

5. You can see no changed items on this panel. To check whether the change history feature is enabled, make a flow on this workspace.

If you are a regular user of Git or GitHub, you should be able to understand the meaning and role of each item just by looking at the structure of this panel. If there is a change in the file structure or contents under the project, the target file will be displayed in the **Local Changes** area. When you move the change to the commit stage (that is, when you add it), the display of the target file moves to the **Changes to commit** area. If you enter a commit message and complete the commit, the version will be incremented by one.

This is exactly the same as what the Git client does.

6. Create a flow and keep it simple. You can make any flow of your choice, for example, here I have used an **inject** node and a **debug** node. Place these two nodes, wire them, and then click the **Deploy** button:

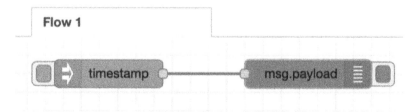

Figure 8.4 – Sample flow to check the project history feature

Following deployment of this flow, you can see the flow.json file in the **Local Changes** area. This means that a flow consisting of an **inject** node and a **debug** node has been added (deployed) on the flow editor, and the flow.json file, which is the configuration file for this entire flow, has been updated. As a result, flow.json has been recognized as a file to be changed in Git management:

Figure 8.5 – Node-RED recognizes that flow.json has been changed

7.   Now, let's follow Git etiquette and proceed with the process. First, put the changed file on the commit stage. This is the git add command of Git.

8.   Click the **Stage all changes** button at the top right of the **Local Changes** area:

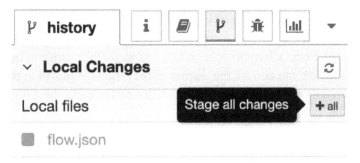

Figure 8.6 – Clicking the Stage all changes button to add the file

You can see that the flow.json file has moved from the **Local Changes** area to the **commit** area.

9.   Next, let's commit the changes in flow.json. Click the **commit** button at the top right of the **Changes to commit** area. This is exactly the equivalent of Git's git commit command:

Figure 8.7 – Clicking the commit button to commit the file

10. After clicking the **commit** button, the commit comment window will be opened. Please enter a commit comment here and then click the **Commit** button:

Figure 8.8 – Clicking the Commit button to complete the commit process

11. The commit is now complete. Finally, let's check the **Commit History** area. You can see that a new version has been created as a change history:

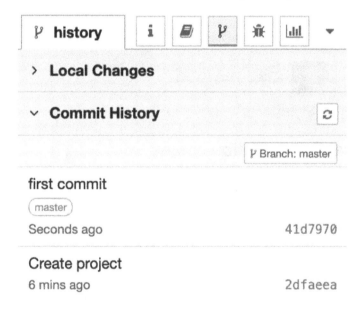

Figure 8.9 – New history has been added

As you learned, after creating your project, it will be possible to use the Node-RED editor the same as usual.

Now, let's add a new user interface to the Node-RED flow editor for project functionality.

Accessing project settings

The project you are working on will appear at the top of the right-hand pane. Next to the project name, there is also a **Show project settings** button:

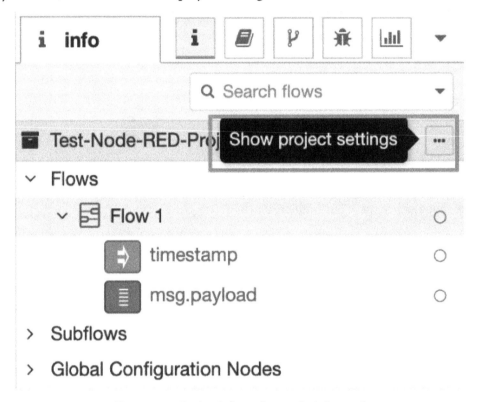

Figure 8.10 – Project information on the info panel

You can also access the **Project Settings** screen from the **Projects | Project Settings** option under the main menu:

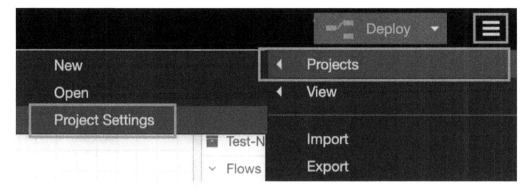

Figure 8.11 – How to access Project Settings via the main menu

When the **Project Settings** panel is shown, you will see that it has three tabs for each setting:

- **Project**: Editing the README.md file of this project
- **Dependencies**: Managing the list of nodes for your project
- **Settings**: Managing the project settings and the remote repositories:

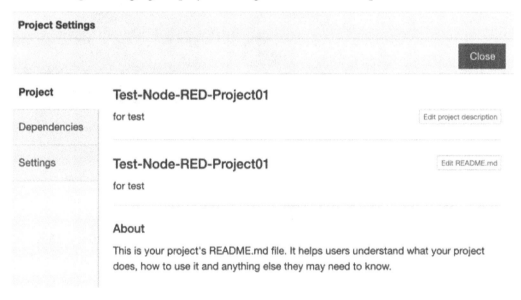

Figure 8.12 – The Project Settings panel

If you want to check and modify the Git settings, you can access the settings panel via the main menu:

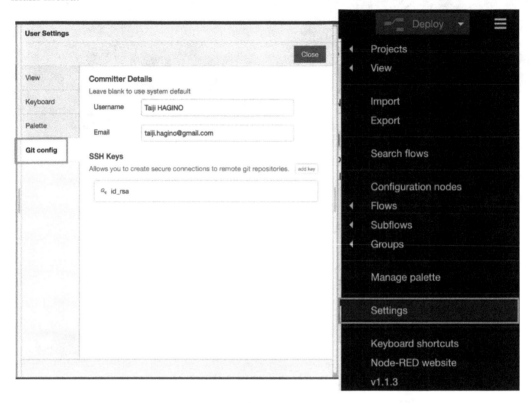

Figure 8.13 – Git config on the User Settings panel

Now you know how to version control in your local environment. The next step entails understanding how to connect a remote repository such as a GitHub service.

Connecting a remote repository

Now, let's learn how to connect Node-RED to a remote repository such as GitHub. Here, we will use the GitHub service for the remote repository. This is like connecting local Git and remote GitHub via Node-RED. This is nothing special. It is familiar to people who use Git/GitHub on a regular basis, but it's very similar to the situation where GitHub is used as a client tool. It is very easy for you to manage the version with Node-RED.

Create a remote repository of your Node-RED project on GitHub with the help of the following steps:

1. First, go to your GitHub account and create a repository.

    It's a good idea to use a project name similar to your local repository. We won't go into details of how to use GitHub here, but since it is a service that can be used intuitively, I believe that anyone can use it without any problems:

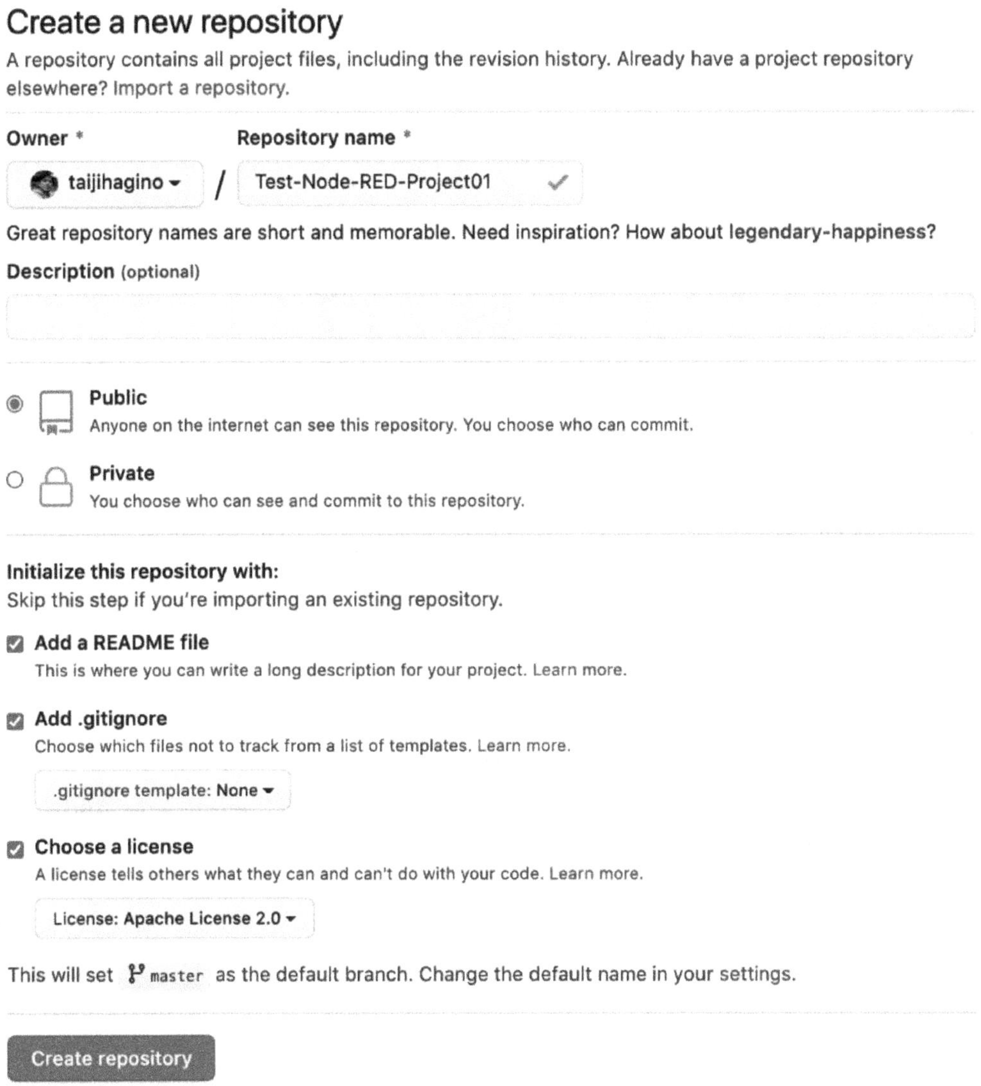

Create a new repository

A repository contains all project files, including the revision history. Already have a project repository elsewhere? Import a repository.

Owner *   Repository name *

🔵 taijihagino ▾   /   Test-Node-RED-Project01   ✓

Great repository names are short and memorable. Need inspiration? How about legendary-happiness?

Description (optional)

⦿ ▢ **Public**
         Anyone on the internet can see this repository. You choose who can commit.

○ 🔒 **Private**
         You choose who can see and commit to this repository.

Initialize this repository with:
Skip this step if you're importing an existing repository.

☑ **Add a README file**
    This is where you can write a long description for your project. Learn more.

☑ **Add .gitignore**
    Choose which files not to track from a list of templates. Learn more.

    .gitignore template: None ▾

☑ **Choose a license**
    A license tells others what they can and can't do with your code. Learn more.

    License: Apache License 2.0 ▾

This will set ⌥ master as the default branch. Change the default name in your settings.

Create repository

Figure 8.14 – Creating a repository on your GitHub

2. Configure the project settings of your Node-RED. To do this, return to the Node-RED flow editor and then go to **Project Settings** to connect the local and remote repositories. When the **Project Settings** panel is opened, click the **add remote** button to configure the remote repository information:

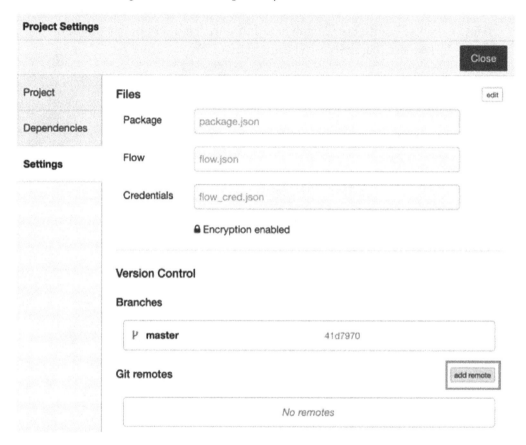

Figure 8.15 – Clicking the add remote button on the Project Settings panel

3. Please enter the repository URL you created on GitHub and then click the **Add remote** button:

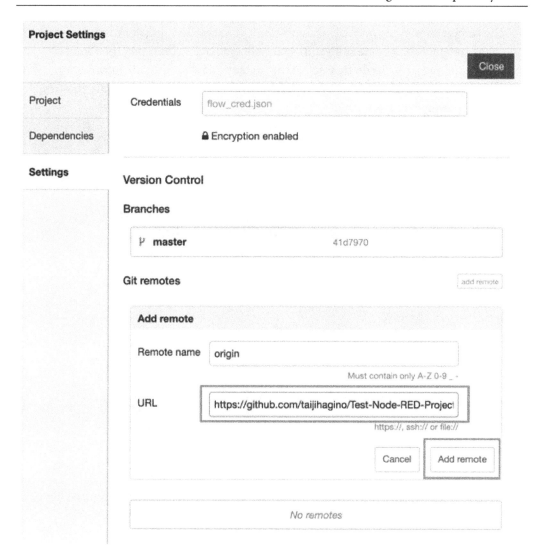

Figure 8.16 – Setting your GitHub repository's URL

4. Click the **Close** button at the top right of the settings panel to complete this configuration.

5. Next, merge the repositories.

The remote repository on GitHub is now linked to the Git repository in your local environment. But they are not yet in sync. All you have to do is pull the remote locally and merge it. To do this, select the **history** panel on the side information menu, and then click the **Manage remote branch** button on the **Commit History** panel to connect to your remote repository:

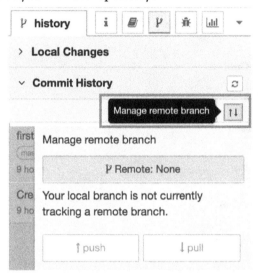

Figure 8.17 – Setting your GitHub repository's URL

6. Select your remote branch. Usually, the **origin/master** branch is selected:

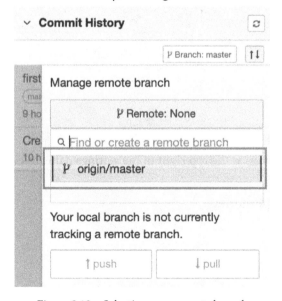

Figure 8.18 – Selecting your remote branch

Here, there is a difference between remote and local because we have already created the flow locally and versioned it with local Git. In this case, you need to pull the remote content locally before you can push the local content to the remote.

7. Click the **pull** button:

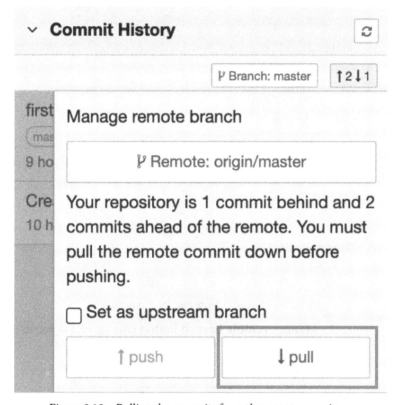

Figure 8.19 – Pulling the commits from the remote repository

A message indicating a conflict will be displayed en route, but proceed with the merge as it is. During the merge, you will be asked whether you want to apply the remote changes or the local changes. In that case, apply the changes on the local side to complete the merge.

Following the operation, you will see that your local branch has been merged with your remote branch on the **Commit History** panel:

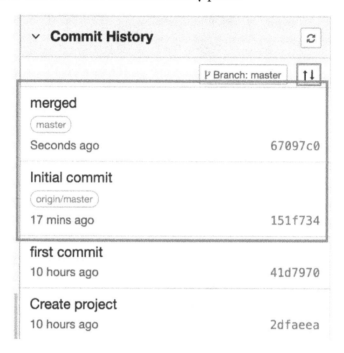

Figure 8.20 – Merged remote and local repositories

8. After this, select the **Manage remote branch** button (the up and down arrows):

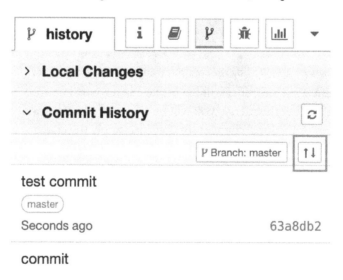

Figure 8.21 – Clicking the Manage remote branch button

9. Select the branch you want to push and then click the **push** button to send (push) these changes to a remote repository:

Figure 8.22 – Sending the changes to the remote repository

Congratulations! Now you have learned how to use the project feature on Node-RED and you can also connect a remote repository from your local repository of Node-RED.

Summary

In this chapter, you learned how to enable the project feature of Node-RED and integrate local version control using Git with a remote repository created on GitHub. This will be very useful when you develop a team using Node-RED in the future.

In the next chapter, we will use this project feature to clone the repository of a to-do application locally. By studying this chapter and the next chapter together, you should have a greater in-depth understanding of the project feature.

Section 3: Practical Matters

In this section, readers will master making realistic and usable applications with Node-RED. The actual application in Node-RED passes the data by separately performing the detailed processing of Node.js. After all the hands-on tutorials in this section, you will have mastered how to use Node-RED.

In this section, we will cover the following chapters:

9
Creating a ToDo Application with Node-RED

In this chapter, we are going to create a simple ToDo application in Node-RED. This is simple and straightforward and is a good tutorial on creating an application (flow) in Node-RED. We are going to use the project feature explained in the previous chapter, so this chapter will also double as a review of that function.

Let's get started with the following four topics:

- Why you should use Node-RED for web applications
- Creating a database
- How to connect to a database
- Running the application

By the end of this chapter, you will have mastered how to make a simple web application with a database on Node-RED.

Technical requirements

To progress through this chapter, you will need the following:

- Node.js 12.x or above (`https://nodejs.org/`).

- CouchDB 3.x (`https://couchdb.apache.org/`).

- A GitHub account, available from `https://github.com/`.

- The code used in this chapter can be found in `Chapter09` at `https://github.com/PacktPublishing/-Practical-Node-RED-Programming`.

Why you should use Node-RED for web applications

So far, this book has explained that Node-RED is an easy-to-use tool for the **Internet of Things (IoT)**. There are many cases where Node-RED is used as a solution in the IoT field.

However, recently, Node-RED has been recognized as a tool for creating web applications as well as IoT.

I think one of the reasons is that the ideas of *no-code* and *low-code* have become widespread in the world. Nowadays, the number of people who know flow-based programming tools and visual programming tools is increasing, and they are being used in various fields.

It would be natural for Node-RED, which is made with Node.js, to be used for web applications.

The project function that we learned in the previous chapter, in collaboration with Git/GitHub, may also be a part of the flow of web application development culture.

In this chapter, we will create a ToDo application that is very suitable as a piece of development for tutorials.

The overall picture of the application to be created is as follows:

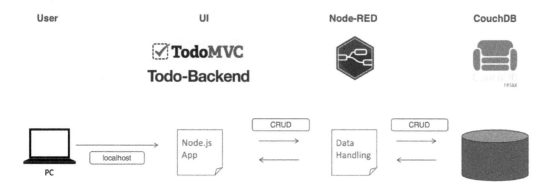

Figure 9.1 – An overview of the application we'll create

Figure 9.1 outlines the application's overview. The application will be accessed from a client PC browser. The user interface of this application is made with Node.js frameworks, **TodoMVC** and **Todo-Backend**. The data-handling programming is built on Node-RED by connecting CouchDB as the storage for this data.

In this application, the user interface and backend application are not built on Node-RED.

The application is implemented on your localhost directly as a Node.js application. We will cover this in a later step, where we will set it to redirect to the localhost Node.js application when accessing the localhost port where Node-RED is running.

There are two frameworks used for this application that we should be aware of before we move toward the hands-on example. We will make our ToDo application with Node-RED in this hands-on tutorial. The application is implemented via these two Node.js frameworks:

- **TodoMVC**: http://todomvc.com/

Figure 9.2 – TodoMVC

- **Todo-Backend**: https://todobackend.com/

Todo-Backend   Implementations   Contribute

Todo-Backend

a shared example to showcase backend tech stacks

The Todo-Backend project defines a simple web API spec - for managing a todo list. Contributors implement that spec using various tech stacks. Those implementations are cataloged below. A spec runner verifies that each contribution implements the exact same API, by running an automated test suite which defines the API.

The Todo-Backend project was inspired by the TodoMVC project, and some code (specifically the todo client app) was borrowed directly from TodoMVC.

Created and curated by Pete Hodgson.

Figure 9.3 – Todo-Backend

As you can see from the fact that it is possible to create a Node-RED flow by linking web application frameworks, Node-RED works very well with the web applications implemented in Node.js and the frameworks around it. This hands-on tutorial will help you understand why Node-RED is so popular for developing web applications in a no-code/low-code fashion.

Next, we will move to the hands-on steps.

Creating a database

We introduced the big picture of the application in the previous section, but more specifically, this application uses CouchDB for the database. In this hands-on tutorial, we will create an application with Node-RED running on localhost. Therefore, you need to install CouchDB on your own local machine as well.

Let's install it by following these steps:

1. Access the CouchDB website at `https://couchdb.apache.org/` and then click the **DOWNLOAD** button:

Figure 9.4 – Click the DOWNLOAD button

2. Select a file depending on the system running on local machine:

Figure 9.5 – Select file

3. Expand the ZIP file you downloaded and run the application file to start CouchDB once the file has finished downloading:

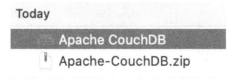

Figure 9.6 – Start CouchDB

4. Running the CouchDB application file launches a browser and opens the CouchDB management console. If it doesn't open automatically, you can also open it manually from the application menu:

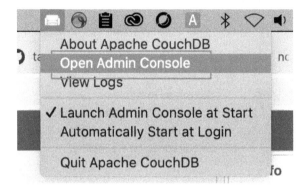

Figure 9.7 – Open the CouchDB admin console

5. In the CouchDB management console, create a new database. Create it with the name todos. No partition is needed. Finally, click the **Create** button to complete:

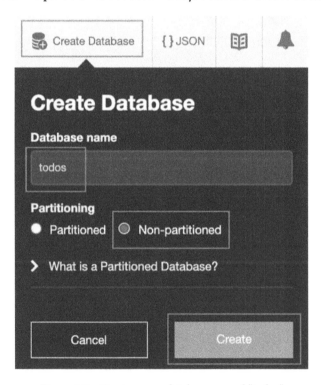

Figure 9.8 – Create a new database named "todos"

You will now be able to see the database named **todos** on your CouchDB admin console:

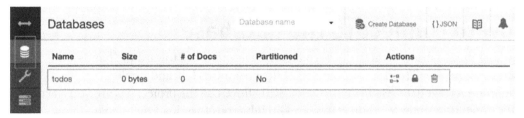

Figure 9.9 – Check the database you created

6. Create an admin user to access this database from your application. To do this, access **User Management** from the side menu of the CouchDB Management Console, select the **Create Server Admin** tab, and set the user account and password.

Here, admin is set as the username and adminpass is set as the password:

Figure 9.10 – Create a server admin user account

This completes all the settings related to CouchDB. Next, let's move on to setting up our Node-RED side.

How to connect to the database

Now that the database has actually been created, we will move toward the hands-on tutorial, where we will clone the Node-RED flow from GitHub, and implement the connection to that database from the Node-RED flow. Use the project feature you learned in the previous chapter to connect to your GitHub repository, load the prepared flow definition file, and implement it on Node-RED in your local environment. Since you have already done this in the previous chapter, it is not necessary to create a new flow this time.

Configuring Node-RED

The first thing you need to do is change the localhost path (URL) of the Node-RED flow editor. Currently, you can access the flow editor at `localhost:1880`, but in order to change the path (URL) of the web application created by this hands-on tutorial to `localhost:1880`, we need to change the path of the flow editor to `localhost:1880/admin`.

This is because you have to move the root path of the Node-RED flow editor to access the Node.js ToDo application running on the same port on your localhost.

To configure Node-RED, follow these steps:

1. Open the settings file (`~/.node-red/settings.js`).

2. Find the `httpAdminRoot` setting in the `settings.js` file you opened.

    This changes the path you access the Node-RED flow editor on. By default it uses the root path `/`, however, we want to use that for our application, so we can use this setting to move the editor. It is commented out by default, so uncomment it by removing the `//` at the start of the line:

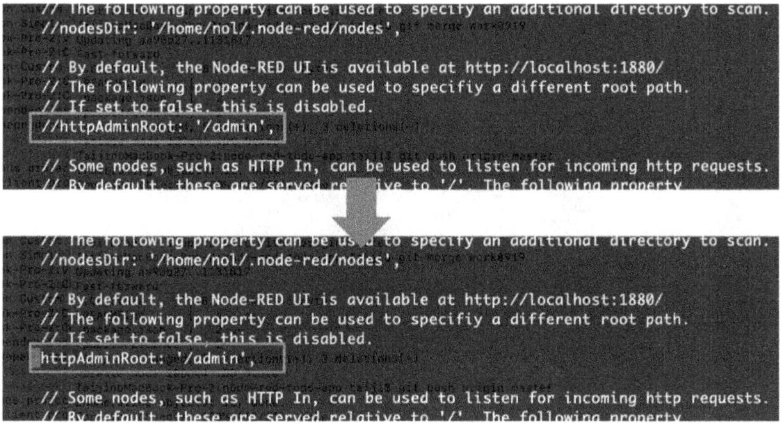

Figure 9.11 – Uncomment httpAdminRoot to enable the flow editor path

3. You have now moved the flow editor to /admin. Restart Node-RED on your local
    machine and access the http://localhost:1880/admin URL to run your
    Node-RED flow editor.

Next, let's clone the project.

Cloning the Node-RED Project

This hands-on tutorial provides an example of a Node-RED project for you to use. Before
cloning it into your local Node-RED instance, you should first fork the project so you have
your own copy of it to use.

After forking it, you need to clone the project into your Node-RED instance.

To clone your project, follow these steps:

1. Open the example project at https://github.com/taijihagino/node-
    red-todo-app.
2. Click the **fork** button to create your own copy of the repository.
3. Copy the URL of the repository you forked.
4. Access the Node-RED editor via http://127.0.0.1:1880/admin/.

5. Click the **Clone Repository** button in the **Projects Welcome** screen. If you've already closed that screen, you can reopen it with **Projects | New** from the main menu:

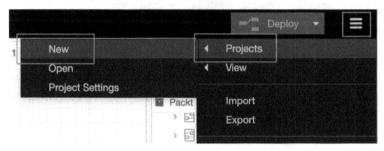

Figure 9.12 – Click New under the Projects menu to clone the repo

6. On the **Projects** screen, provide your repository URL, your username, and password. These are used when committing changes to the project. If your local Git client is already configured, it will pick those values. It is fine to leave the **Credentials encryption key** field blank:

Figure 9.13 – Provide your GitHub repository information

7. This will clone the repository into a new local project and start running it. In the workspace, you can see flows that implement each part of the application's REST API.

You will see some errors on all of the **cloudant** nodes, but the reasons for these errors come from the connection settings. These settings will be made in later steps so it is not a problem for now:

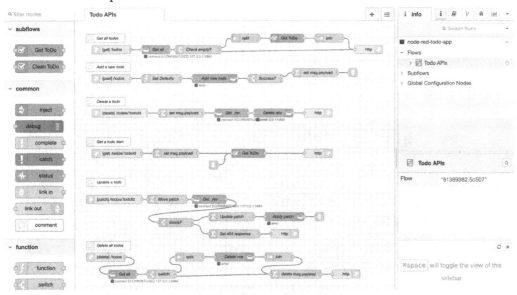

Figure 9.14 – The flow overview you cloned

8. The project also includes some static resources that need to be served by the runtime. To do this, some changes to how you access this web application need to be made in your settings file.

First, you must locate your newly-cloned project on the local filesystem. It will be in `<node-red root>/projects/<name-of-project>`. Within that folder, you will find a folder named `public`. This contains the static resources for the project of this ToDo application, such as the following, for example:

```
/Users/taiji/.node-red/projects/node-red-todo-app
```

The following image is an example of this. Please use it as a reference when checking your own file path:

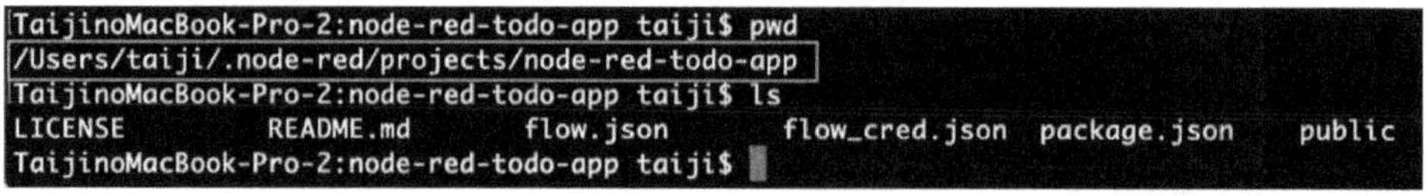

Figure 9.15 – The ToDo application project folder

9. Edit your settings file (~/.node-red/settings.js) and find the httpStatic property in this file. Uncomment it by removing the // at the start of the line and set its value using the absolute path to the public folder. The path in the following image is just an example; please replace it with your path:

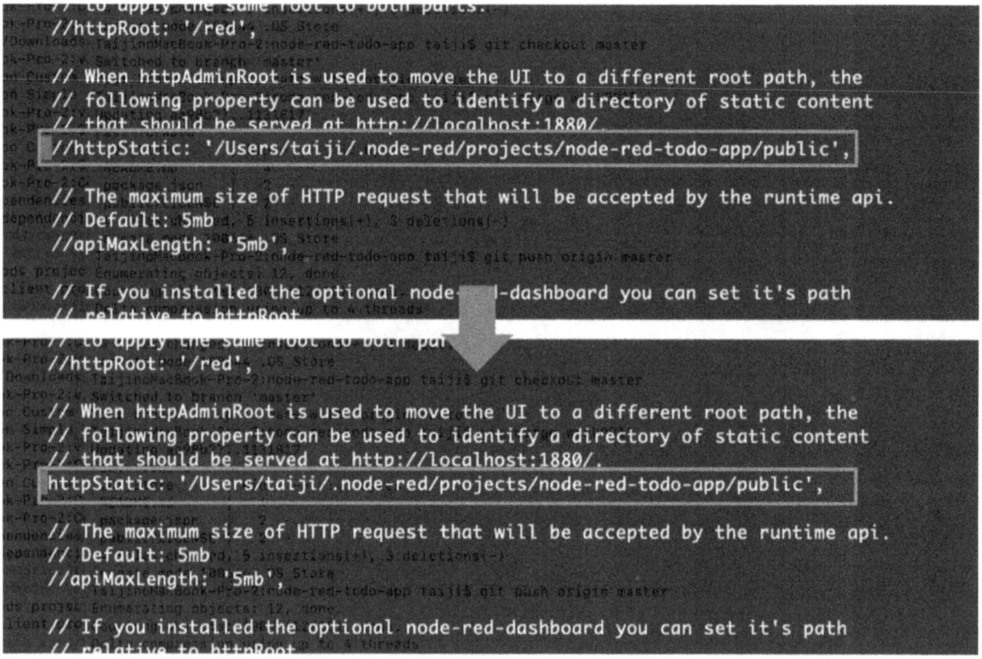

Figure 9.16 – Uncomment httpStatic and set your application project path

10. Restart Node-RED.

By restarting Node-RED, the changed settings.js contents will be reloaded and applied.

Next, let's configure the Node-RED and CouchDB connection.

Configuring the Node-RED and CouchDB connection

As you know, we are using a **cloudant** node to connect to CouchDB, correct?

Cloudant is a JSON database based on Apache CouchDB. Cloudant has CouchDB-style replication and synchronization capabilities, so you can connect to CouchDB using the **cloudant** node provided by Node-RED.

As mentioned earlier, the **cloudant** node on Node-RED is experiencing an error. This is because the connection information to CouchDB on your local system is not set correctly when cloned from GitHub.

Here, we will correct the settings of the **cloudant** node on Node-RED.

Now, carry out the settings according to the following steps:

1. Double-click any **cloudant** node to open the settings screen. If you set one of the **cloudant** nodes there, the settings of all **cloudant** nodes on the same flow will be updated, so it doesn't matter which **cloudant** node you choose:

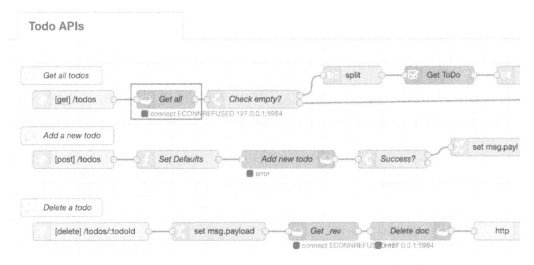

Figure 9.17 – Open the settings screen with a double-click on any cloudant node

2. Click the **pencil mark** button on the right side of **Server** on the **cloudant** node
    settings screen to open the connection information settings screen for CouchDB:

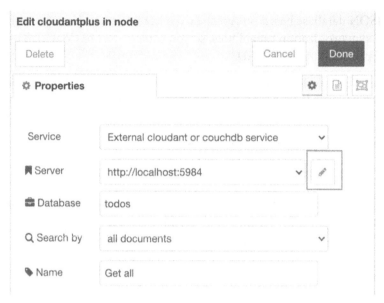

Figure 9.18 – Click the pencil mark button

3. When the connection information settings screen for CouchDB opens, go to
    Host and set it to `http://localhost:5984` (if you have CouchDB installed
    on a different port, replace it as appropriate) and set the **Username** to the server
    admin user of CouchDB that you set earlier. For **Password**, enter the server admin
    password.

4. After entering all of this, click the **Update** button on the upper right to return to the previous screen:

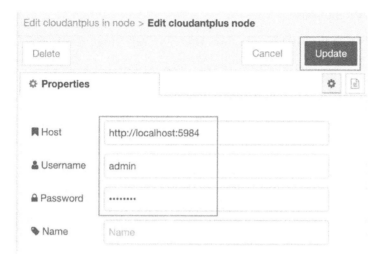

Figure 9.19 – Set your CouchDB URL and server admin user/password

5. Click the **Done** button and return to the workspace of your Node-RED flow editor. You will see a message reading **connected** on all of the **cloudant** nodes next to a green square:

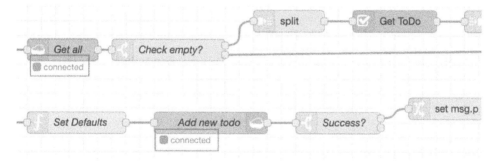

Figure 9.20 – Check that all of the cloudant nodes are error-free

Perfect, you have succeeded in configuring the settings for starting the ToDo application in Node-RED. Next, let's run this ToDo application.

Running the application

If everything is working, you should be able to open `http://localhost:1880` in the browser and see the application.

Now, let's confirm that the ToDo application works by following these steps:

1. Access `http://localhost:1880` to open your ToDo application.

    If you get the Node-RED flow editor when you open `localhost:1880`, the setting `httpAdminRoot` is not enabled, so please check your `settings.js` file again.

    When you access this URL, the following screen should be displayed:

Figure 9.21 – Open your ToDo application

2. Any ToDo item is fine for this test, so enter any words as a sample task. Here, I typed `Report my tasks`:

Figure 9.22 – Enter a sample ToDo item

3. If you press the *Enter* key while entering a value in the text box, that value will be registered as a ToDo item. In the following screenshot, we can see that it looks like it has been registered in the application:

Figure 9.23 – The ToDo item you entered has been registered

Let's check if the ToDo item that showed as registered on the screen is registered in the database.

4. Open the CouchDB admin console.

    If you forget how to open it, you can open it with the **Open Admin Console** option from the CouchDB application menu. If you reopen the management console, or if the time has passed, you may be asked to log in. In that case, log in with the server admin username and password you set.

5. Select **Database** option in the side menu, and click **todos**. You will see the record you registered on your ToDo application. Click the record to show more details:

Figure 9.24 – Check the record on your todos database

6. You will see the detail of the record you selected. The data is the exact item you registered via the ToDo application, that is, **Report my tasks**:

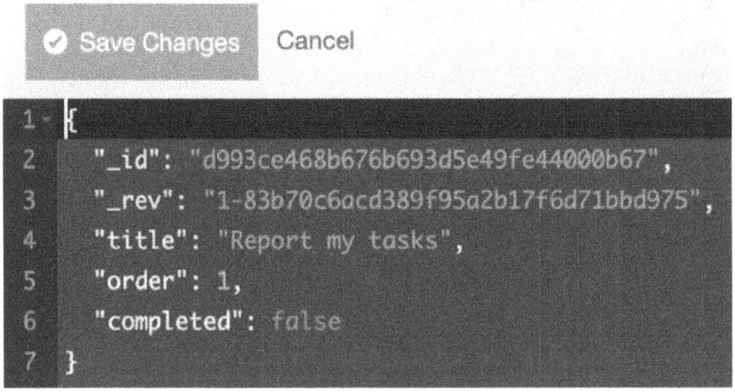

Figure 9.25 – Check the result

Congratulations! This completes the hands-on tutorial for cloning a ToDo application from GitHub and implementing it in Node-RED.

The point of this tutorial was to use the project function of Node-RED to clone and execute the application project from the GitHub repository.

This hands-on tutorial helped us learn that we don't necessarily have to implement user interfaces and server-side business logic in web applications made with Node-RED. We saw how one of the features of Node-RED is that the user interfaces and server-side business logic of the web application that we built are located outside of Node-RED, while only data handling functionalities such as accessing the database are done internally by Node-RED.

The GitHub repository we used contains two things, that is, Node-RED flow, which handles data, and the ToDo application, which runs outside Node-RED. The point here was to use the project function of Node-RED to clone and execute the application project from the GitHub repository.

Summary

In this chapter, in the form of a hands-on tutorial, we experienced how to actually run a web application on Node-RED using the project feature. Of course, this is just one way to create a web application (including a UI, using a template node, and so on) on Node-RED. However, remembering this pattern will definitely be useful for your future development tasks.

In the next chapter, we will look at a hands-on scenario where we will be sending sensor data from an edge device to the server side (cloud) with Node-RED.

10
Handling Sensor Data on the Raspberry Pi

In this chapter, we will learn how the processing of data from an edge device takes place in the **Internet of Things** (**IoT**) using Node-RED. We will not only cover data handling but also sending data to a server application from an edge device. For the device, I would like to use a Raspberry Pi. After completing the tutorials in this chapter, you will be able to handle sensor data acquired by edge devices.

Let's get started with the following four topics:

- Getting sensor data from the sensor module on the Raspberry Pi
- Learning the MQTT protocol and using an MQTT node
- Connecting to an MQTT broker
- Checking the status of data on localhost

Technical requirements

To progress in this chapter, you will need the following:

- A Raspberry Pi, available from `https://www.raspberrypi.org/`
- The code used in this chapter can be found in `Chapter10` folder at `https://github.com/PacktPublishing/-Practical-Node-RED-Programming`

Getting sensor data from the sensor module on the Raspberry Pi

In this chapter, we will learn how to handle the data acquired from the sensor device with Node-RED on the Raspberry Pi and publish the data to an MQTT broker.

For the sensor device, we will use the temperature/humidity sensor used in *Chapter 5, Implementing Node-RED Locally*. See each step in *Chapter 5, Implementing Node-RED Locally*, for details about connectivity and how to enable the sensor device on the Raspberry Pi.

Prepare to connect your temperature/humidity sensor to your Raspberry Pi. This is the edge device. You have already purchased and configured your edge device in *Chapter 5, Implementing Node-RED Locally*. Light sensors are not used in this chapter:

- Edge device: **Raspberry Pi 3** (`https://www.raspberrypi.org/`)
- Sensor module: **Grove Base HAT for Raspberry Pi, Grove Temperature and Humidity Sensor (SHT31)** (`https://www.seeedstudio.com/Grove-Base-Hat-for-Raspberry-Pi.html`, `https://www.seeedstudio.com/Grove-Temperature-Humidity-Sensor-SHT31.html`)

Preparing the devices

Please prepare the device to gather the temperature/humidity sensor data on your Raspberry Pi as follows:

1. Connect the sensor module to your Raspberry Pi.

   When all the devices are ready, connect the Raspberry Pi and Grove Base HAT, and connect the Grove Temperature and Humidity Sensor (SHT31) to the I2C port (any I2C port is OK):

Figure 10.1 – Connecting the temperature/humidity sensor to your Raspberry Pi

2.   Connect your Raspberry Pi to the internet.

We will go on to connect to the server side from the Raspberry Pi, so please ensure that you are connected to the internet via Wi-Fi. Of course, you can also access the internet by connecting to a modem using a LAN cable. The Raspberry Pi has a LAN cable port by default, so all you have to do is plug in the LAN cable:

Figure 10.2 – Connecting your Raspberry Pi to the internet

And that's all we need to proceed. Next, we will see how to get the data from the sensor node.

Checking Node-RED to get data from the sensor device

As you have already learned in *Chapter 5*, *Implementing Node-RED Locally*, it should be easy to get the data from the Grove Base temperature/humidity sensor module.

The following are the steps to get the data from the sensor node:

1. Make a simple flow to get the data. Select three nodes, that is, an **inject** node, a **grove-temperature-humidity-sensor-sht3x** node, and a **debug** node, from the palette on the left side of the flow editor and drag and drop them into the workspace to place them.

2. After placing them, please wire them sequentially as shown in the following diagram:

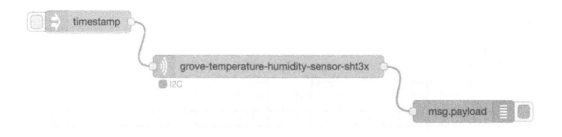

Figure 10.3 – Placing and wiring nodes

3. Check the settings of the **grove-temperature-humidity-sensor-sht3x** node. To check the settings, double-click the **grove-temperature-humidity-sensor-sht3x** node to bring up the settings screen.

    There are no values or items to be set on this settings screen. You just need to make sure that the port is indicated as **I2C**. After checking, close the settings screen.

    Make sure you see a blue square icon and the text **I2C** underneath the **grove-temperature-humidity-sensor-sht3x** node. This indicates that the Grove Base temperature/humidity sensor module is successfully connected to your Raspberry Pi. If the color of this icon turns red, it means that the module is not properly connected to the **I2C** port, so please reconnect the hardware correctly:

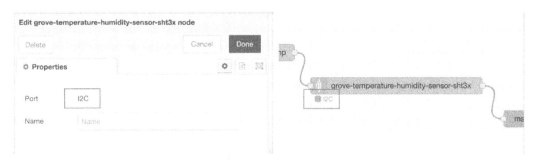

Figure 10.4 – Checking the port is set as I2C

4. Execute the flow and check the results by clicking the **Deploy** button in the upper right corner of the flow editor to complete the deployment.

5. Once the deployment is successful, click the switch on the **inject** node to start the flow:

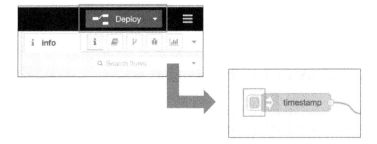

Figure 10.5 – Deploy and click the button on the inject node

It has worked successfully if you can confirm that the values of the sensor data collected are displayed in JSON in the **debug** tab of the flow editor. This way, data can be obtained from the sensor module:

Figure 10.6 – Making sure that the data is visible from the sensor module

Now we know that Node-RED on the Raspberry Pi can handle sensor data. Let's learn the process of publishing this data to an MQTT broker.

Learning the MQTT protocol and using an MQTT node

Now that the sensor data has been successfully acquired, let's send that data to the server.

We usually select a protocol suitable for the content being transmitted; for example, when exchanging mail, we use SMTP. Currently, HTTP is used as a general-purpose protocol on the internet.

For example, HTTP is used for various communications on the internet, such as displaying web pages in a browser and exchanging data between servers. HTTP is a protocol created for exchanging content on the internet. In many cases, network devices such as routers and firewalls on the internet are set to allow HTTP communication to be used for various purposes, and HTTP is compatible with the internet.

In the IoT world, MQTT is often used as a general protocol instead of HTTP. This means that the MQTT protocol is the standard of the IoT world, just as the HTTP protocol is the standard of the web world.

MQTT (short for **MQ Telemetry Transport**) is a communication protocol that was first created by IBM and Eurotech in 1999. In 2013, standardization of this protocol was promoted by an international standardization organization called OASIS.

MQTT is intended to be used over TCP/IP. In short, it specializes in **machine-to-machine (M2M)** communication over the internet, and communication between machines and other resources on the internet. The *machines* referred to here are microcomputer boards, such as PCs and small Linux boards (including the Raspberry Pi).

M2M has evolved over the years since 1999, the word **IoT** has appeared, and MQTT is now very often adopted when conventional machines communicate via the internet. Therefore, MQTT is the best protocol for IoT. One of the reasons that MQTT is important is that it offers a lightweight protocol to handle data in narrowband networks and on low-performance devices:

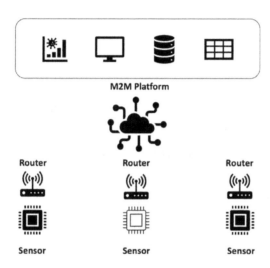

Figure 10.7 – Conceptual diagram of typical M2M communication

From the preceding information, you can see why the MQTT protocol is used in IoT. Now let's think about how Node-RED can transmit data using the MQTT protocol.

Node-RED provides the following two MQTT related nodes by default:

- **mqtt in**: The **mqtt in** node connects to the MQTT broker and subscribes to messages on the specified topic.

- **mqtt out**: The **mqtt out** node connects to the MQTT broker and publishes messages:

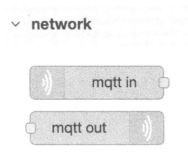

Figure 10.8 – An mqtt in node and mqtt out node

You can find these under the **network** category on the side panel of the Node-RED flow editor.

If you want to set the server address and topic for the MQTT broker and use publish and subscribe, it is fine to use these two nodes.

Let's now try to send the sensor data to a local MQTT broker.

Connecting to an MQTT broker

Now, let's send the sensor data on the Raspberry Pi to an MQTT broker via Node-RED. Here we will use the popular MQTT broker **Mosquitto**. In this chapter, we will go as far as preparing the device to send the device data to the server. The task of actually receiving and processing data on the server side will be demonstrated in a hands-on example in the next chapter. Therefore, here we will use Mosquitto just for checking the data transmission is performed correctly.

Mosquitto

Mosquitto is released under the open source BSD license and provides broker functionality for MQTT V3.1/v3.1.1.

It works on major Linux distributions such as RedHat Enterprise Linux, CentOS, Ubuntu, and OpenSUSE, as well as Windows. It also works on small computers such as the Raspberry Pi.

In this chapter, we will verify that the sensor data of the edge device can be sent via an MQTT broker to the localhost of the Raspberry Pi. This is very easy. I am confident that if we can send the data to MQTT broker in this way, we will be able to see the sensor data of the edge device immediately on the server side.

The following is a general configuration diagram showing an example use of Mosquitto:

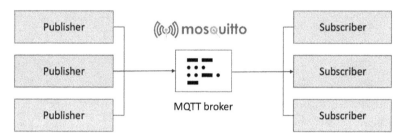

Figure 10.9 – Mosquitto overview

In this chapter, we will implement the Node-RED flow from the edge device to send data to Mosquitto on your Raspberry Pi. Data visualization using IBM Cloud will be implemented in the next chapter.

> **Important note**
>
> Mosquitto is a very important and useful tool and is a platform for implementing the IoT mechanism in Node-RED. Developing a deeper understanding of Mosquitto will help you to make Node-RED more widely available.
>
> You can learn more about the Mosquitto at `https://mosquitto.org/`.

Now, let's prepare Mosquitto on your Raspberry Pi.

Preparing Mosquitto on your Raspberry Pi

In this section, we will enable Mosquitto so that it can run on the Raspberry Pi. The flow is simple. Just install Mosquitto and start the service. Follow these steps on your Raspberry Pi to prepare:

1. To install Mosquitto, execute this command on the terminal:

```
$ sudo apt install mosquitto
```

2. To start the Mosquitto service, execute this command on the terminal:

```
sudo systemctl start mosquitto
```

   After starting, you can check the status of the Mosquitto service with the following command:

```
sudo systemctl status mosquitto
```

   This is how it looks in the terminal:

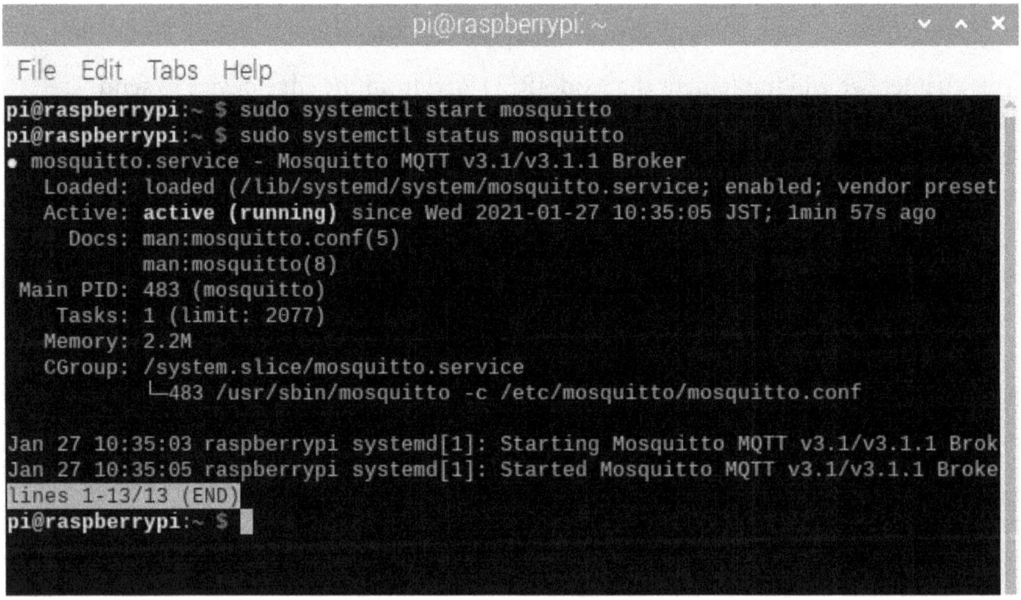

Figure 10.10 – Mosquitto running status

3. To install the Mosquitto client tool, execute this command on the terminal:

    ```
    $ sudo apt install mosquitto-clients
    ```

4. To check the publish and subscribe functionality, run **Subscriber** on your Raspberry Pi with the following command. Here we set `packt` as the **topic**:

    ```
    $ sudo apt install mosquitto-clients
    $ mosquitto_sub -d -t packt
    ```

    This is how it looks in the terminal:

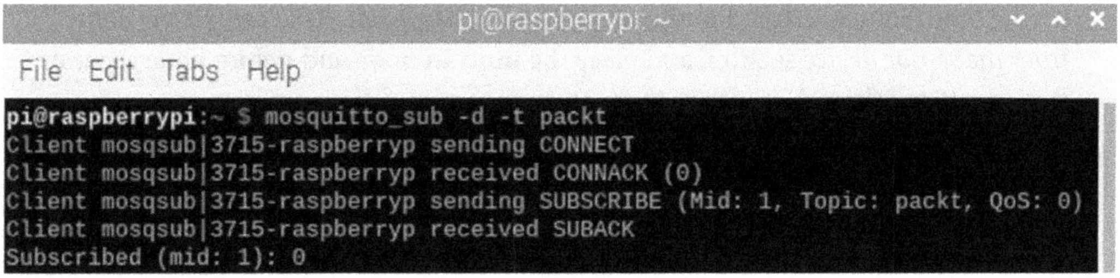

Figure 10.11 – Start subscribing to Mosquitto with the topic packt

5. Publish some text to this broker with the following command on another terminal:

    ```
    $ mosquitto_pub -d -t packt -m "Hello Packt!"
    ```

    This is how it looks in the terminal:

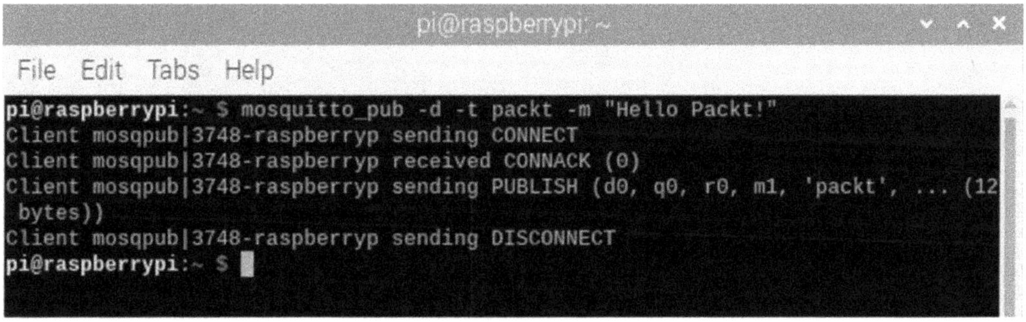

Figure 10.12 – Publishing a message to Mosquitto with the topic packt

You will see the message you published on the terminal subscribing.

You are now ready to use Mosquitto. Next, we will implement Pub/Sub on Node-RED on your Raspberry Pi.

Making a flow to get sensor data and send it to the MQTT broker

Now, launch the Node-RED flow editor on your Raspberry Pi and follow these steps to create a flow:

1. Place the **mqtt out** node after the **grove-temperature-humidity-sensor-sht3x** node on the flow that you created in the previous *Checking Node-RED can get the data from the sensor device* section, and place the **mqtt in** node and **debug** node separate from **mqtt out flow**. Please wire them as shown in the following figure:

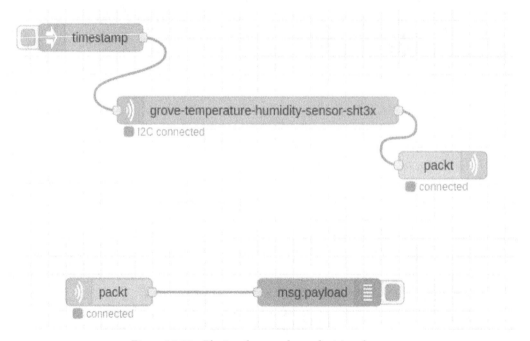

Figure 10.13 – Placing these nodes and wiring them

2. Edit the **mqtt out** node by double-clicking on it and set the values in the **Properties** tab as follows to connect to the **Mosquitto** MQTT broker you have run:

- Server: `localhost`
- Port: `1883`

    *It is possible to edit the **Server** and **Port** values by clicking the *pencil* icon.

- **Topic**: `packt`
- **Qos**: `1`
- **Retain**: `true`

This is how the settings window should look:

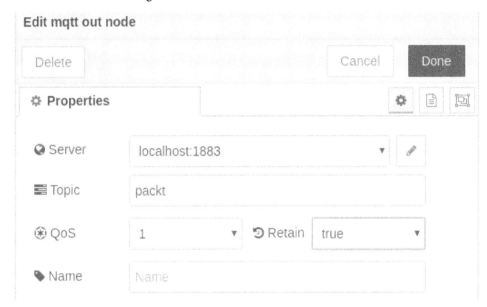

Figure 10.14 – Setting the properties of the mqtt out node

3. Edit the **mqtt in** node by double-clicking it so the settings window appears. Set the values on the **Properties** tab as follows to subscribe to the topic from the **Mosquitto** MQTT broker you have run:

- **Server**: `localhost`
- **Port**: `1883`

   *It is possible to edit the **Server** and **Port** values by clicking the *pencil* icon.

- **Topic**: `packt`
- **Qos**: `1`

- **Output: auto-detect (string or buffer)**

This is how the settings window should look:

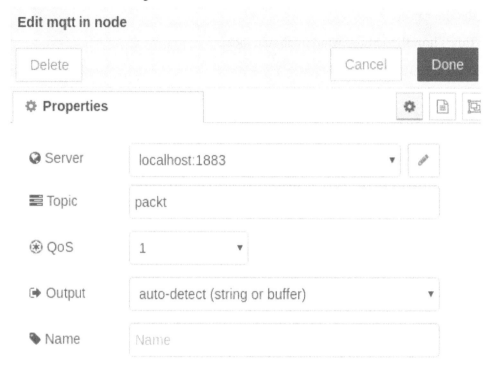

Figure 10.15 – Setting the properties of the mqtt in node

And with that, we have completed making the flow to subscribe to and publish the topic `packt` via the **Mosquitto** MQTT broker on your Raspberry Pi localhost. Next, we will check the status of our data on localhost.

Checking the status of data on the localhost

In this section, we will check whether the sensor data sent from your Raspberry Pi can be received by Mosquitto via Node-RED on your Raspberry Pi with the following steps:

1. Run the flow you created in the previous section on the Node-RED instance on your Raspberry Pi.

2. Click the switch of the **inject** node to run this flow and publish the Grove temperature and humidity sensor data:

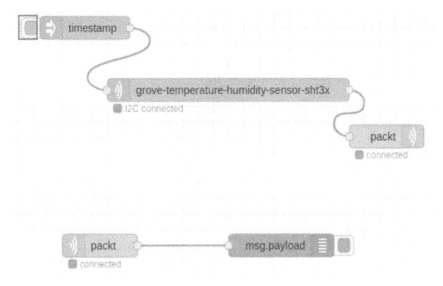

Figure 10.16 – Run the flow to publish the data

3. Check that the data was subscribed.

    There are currently two flows in this Node-RED instance. One is the flow of publishing data to the Mosquitto MQTT broker, and the other is the flow of subscribing to data from that broker. The subscribed flow is normally in a standby state, so when the data is published, the subscribed data is automatically output to the **debug** tab.

4. Check the **debug** tab. You should see the data you published:

Figure 10.17 – Check the result of the publishing and subscribing\

Congratulations! Now you know how to handle the sensor data acquired by the Raspberry Pi and Grove Base sensor module on the edge device and send it to an MQTT broker.

Summary

In this chapter, in the form of a hands-on tutorial, we experienced how to handle sensor data on an edge device and send it to an MQTT broker. This is one of the ways to create an edge device-side application for IoT with Node-RED.

In the next chapter, we will look at a hands-on example of receiving sensor data and visualizing it on the server side (the cloud) via Node-RED.

11

Visualize Data by Creating a Server-Side Application in the IBM Cloud

In this chapter, we will create a server application to visualize data that has been sent from an edge device in the IoT, using Node-RED. For a server-side application, I would like to use the IBM Cloud here. By following the tutorials in this chapter, you will master how to visualize sensor data on a server application.

Let's get started with the following topics:

- Preparing a public MQTT broker service
- Publishing the data from Node-RED on an edge device
- Subscribing and visualizing data on the cloud-side Node-RED

By the end of this chapter, you will have mastered how to visualize sensor data on cloud platforms.

Technical requirements

To progress in this chapter, you will require the following:

- An IBM Cloud account: `https://cloud.ibm.com/`

- A CloudMQTT account: `https://cloudmqtt.com/`

- The code used in this chapter can be found in `Chapter11` folder at `https://github.com/PacktPublishing/-Practical-Node-RED-Programming`.

Preparing a public MQTT broker service

Recall the previous chapter, *Chapter 10, Handling Sensor Data on the Raspberry Pi*. We sent the data of the temperature/humidity sensor, which was connected to the edge device (Raspberry Pi), to the cloud and confirmed that the data could be observed on the cloud side.

In the previous chapter, we checked how to operate an MQTT broker using a service called **Mosquitto**. This was in order to focus on *sending data from edge devices* to an MQTT broker.

However, this was a mechanism where everything was done locally on the Raspberry Pi. Essentially, when trying to implement an IoT mechanism, MQTT brokers should be in a public location and accessible from anywhere via the internet.

It is possible to host your own **Mosquitto** MQTT broker in the public cloud, but that adds some extra complexity in terms of setting up and maintaining it. There are a number of public MQTT services available that can make it easier to get started.

In this chapter, we will use a service called **CloudMQTT** for the MQTT broker, but you can replace the MQTT broker part with your favorite service. You can also publish your own MQTT broker, such as **Mosquitto**, on IaaS instead of using SaaS:

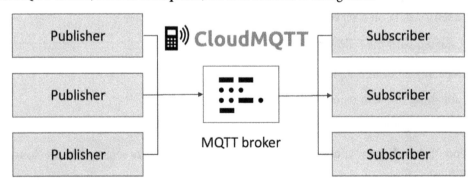

Figure 11.1 – CloudMQTT overview

> **Important note**
>
> An MQTT broker is a server that receives messages from the publisher and sends them to subscribers.
>
> The server that delivers messages in PubSub is called the MQTT broker.
>
> PubSub is an amalgamation of the words *Publisher* and *Subscriber*.
>
> a) A publisher is a person who delivers a message.
>
> b) A subscriber is a person who subscribes to a message.
>
> You can think of it as a server that receives messages from clients and distributes them to clients.
>
> MQTT differs from ordinary socket communication in that it communicates on a one-to-many basis. In other words, it has a mechanism to distribute the message of one client to many people. This system allows us to deliver messages to many people simultaneously in real time.

We will now learn how to prepare for **CloudMQTT**. As mentioned previously, **CloudMQTT** is an MQTT broker published as SaaS. If you don't use **CloudMQTT** and want to use another SaaS MQTT broker or publish an MQTT broker on IaaS, you can skip this step. However, the basic configuration information for using an MQTT broker remains the same, so I believe this step will help you configure any MQTT broker.

Perform the following steps to create an MQTT broker service on **CloudMQTT**:

1. Log in to **CloudMQTT** at `https://cloudmqtt.com/`.

   When you access the website, click the **Log in** button at the top right of the window:

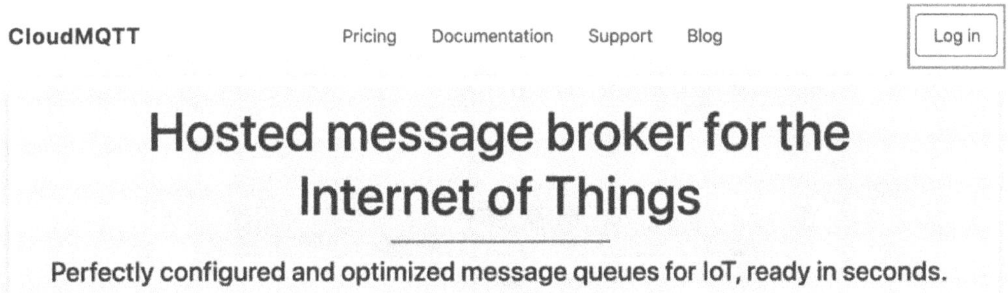

Figure 11.2 – CloudMQTT website

If you already have your CloudMQTT account, log in to your account by entering your email address and password:

Figure 11.3 – Logging in to CloudMQTT

If you don't yet have your account, please create a new account via the **Sign up** button at the bottom of the window:

Figure 11.4 – Creating your account

2. Create an instance.

    After logging in, click the **Create New Instance** button in the top-right corner of the window:

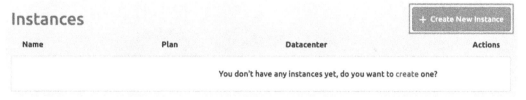

Figure 11.5 – Creating a new instance

3. Select a name and payment plan.

    This name is for your MQTT broker service. You can give it any name you want. I have used `Packt MQTT Broker`.

    Unfortunately, the free plan, **Cute Cat**, is no longer available. So, we will select the cheapest plan, **Humble Hedgehog**, here. This plan costs $5/month.

    It's up to you to use this paid service. If you want to avoid billing, you need to look for a free MQTT broker service.

    After selecting the name and payment plan, click the **Select Region** button:

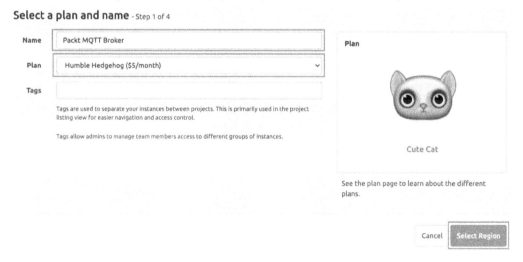

Figure 11.6 – Selecting a name and payment plan

4. Select a region and data center.

    This service is running on **AWS**. So, you can select a region where the data center is placed. You can select any region. Here, we are using **US-East-1**.

5.   After making the selection, click the **Review** button:

Figure 11.7 – Selecting a region and data center

6.   Next, finalize creation of the MQTT broker instance.

Please check the payment plan, service name, service provider, and data center region. After that, click the **Create instance** button to finalize creation of this instance:

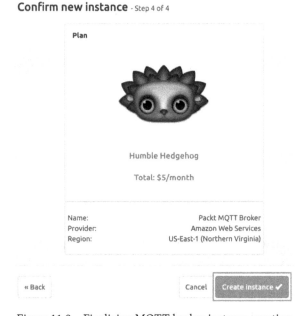

Figure 11.8 – Finalizing MQTT broker instance creation

Publishing the data from Node-RED on an edge device

In this section, we will configure our Raspberry Pi. To get started, launch the Raspberry Pi and open the Node-RED flow editor. This Node-RED flow editor should still have a flow to send the sensor data implemented in *Chapter 10*, *Handling Sensor Data on the Raspberry Pi*, to the server. If you have deleted this flow, or if you have not created it, please re-execute it by referring to *Chapter 10*, *Handling Sensor Data on the Raspberry Pi*. Double-click the **mqtt out** node that makes up the flow to open the settings window. We used **Mosquitto** last time, but this time we will connect to **CloudMQTT**.

Perform the following steps to configure Node-RED on the Raspberry Pi to connect to CloudMQTT:

1. Access the flow you created in *Chapter 10*, *Handling Sensor Data on the Raspberry Pi*.

    In this chapter, we only use a flow with the **mqtt out** node because this scenario is just for sending data to a Raspberry Pi:

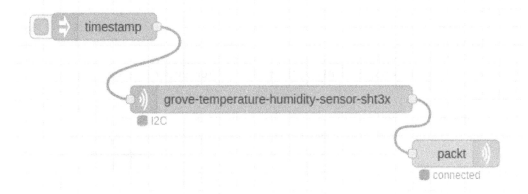

Figure 11.9 – Accessing the flow we created in the previous chapter

2.   Edit the **mqtt out** node.

We now need to edit the connecting configuration. Open the settings window of the **mqtt out** node by double-clicking on it:

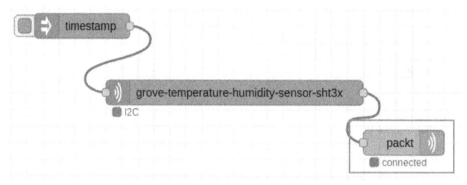

Figure 11.10 – Opening the settings window of the mqtt out node

Set the configuration to connect to CloudMQTT.

Set the **Topic, Qos**, and **Retain** values as follows:

- **Topic**: packt

- **Qos**: 1

- **Retain**: true

3.   Click the **Edit** button (pencil mark) to the right of **Server** to open the credential properties:

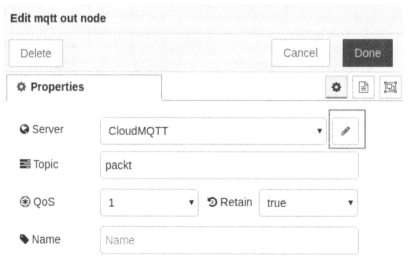

Figure 11.11 – Clicking the Edit button to open the Properties settings

4. On the Server settings panel, select the **Connection** tab and fill in each property as follows:

- **Server**: `driver.cloudmqtt.com`
- **Port**: `18913`

    The other properties in the **Connection** tab are not supposed to be changed and must be kept at their default values.

    You can refer to the following screenshot for the **Connection** tab settings:

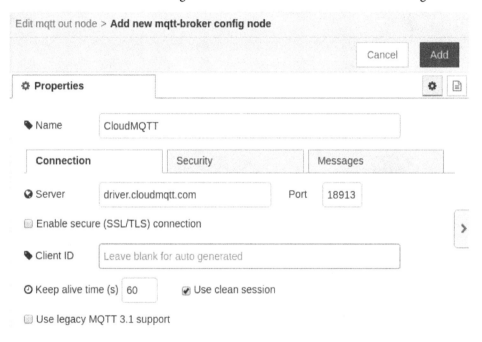

Figure 11.12 – MQTT broker server settings

5. Next, select the **Security** tab to edit the configuration to connect the MQTT broker and fill in each property as follows:

-   **Username**: The user that you got from CloudMQTT.

-   **Password**: The password that you got from CloudMQTT.

    You can refer to the following screenshot for the **Security** tab settings:

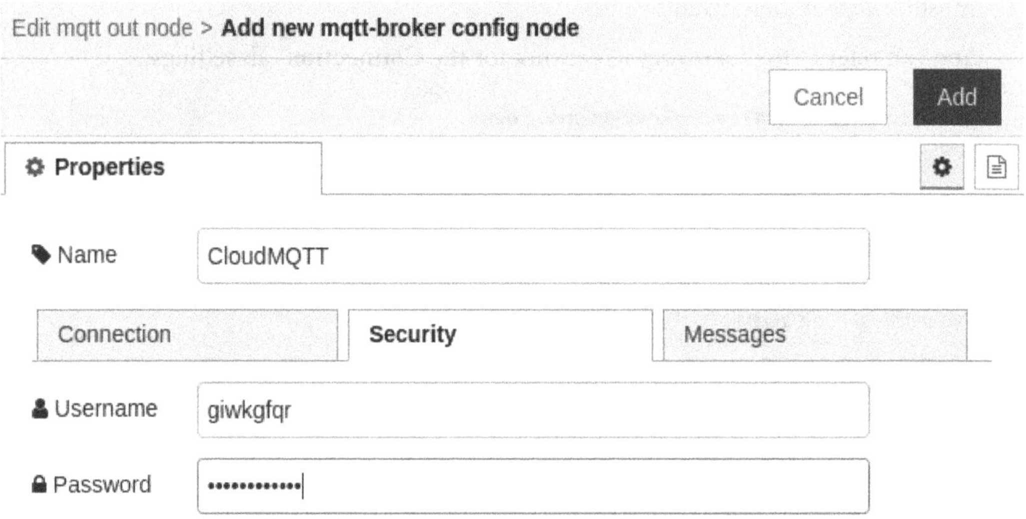

Edit mqtt out node > **Add new mqtt-broker config node**

	Cancel	Add

⚙ **Properties**   ⚙   ▤

🏷 Name   CloudMQTT

Connection	**Security**	Messages

👤 Username   giwkgfqr

🔒 Password   ••••••••••••

Figure 11.13 – MQTT broker user and password settings

You can check these properties at the CloudMQTT admin menu. This menu can be accessed via the Instances list of the CloudMQTT dashboard: `https://customer.cloudmqtt.com/instance`

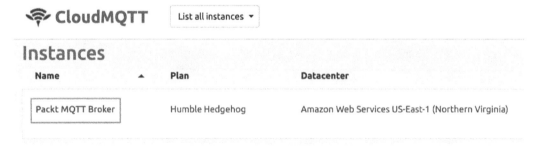

Figure 11.14 – CloudMQTT Instances list

This completes the settings on the Raspberry Pi side. Next, let's set up the Node-RED flow editor so that data can be acquired (subscribed) with Node-RED on the cloud side.

Subscribing and visualizing data on the cloud-side Node-RED

In this section, we will see how to visualize the received data with Node-RED on the cloud side. This uses one of the dashboard nodes as we learned in *Chapter 6, Implementing Node-RED in the Cloud*, but this time, we'll choose Gauge's UI to make it look a little better.

The cloud-side Node-RED used this time runs on the IBM Cloud (PaaS), but CloudMQTT, which created the service as an MQTT broker earlier, is a cloud service that differs from the IBM Cloud.

In this chapter, we will learn that an MQTT broker exists so that it can be accessed from various places, and that both publishers (data distributors) and subscribers (data receivers) can use it without being aware of where it is.

Preparing Node-RED on the IBM Cloud

Now, let's create a Node-RED flow connected to CloudMQTT by performing the following steps. Here, we will use Node-RED on the IBM Cloud. Please note that it is not Node-RED on the Raspberry Pi:

1. Open the Node-RED flow editor, log in to your IBM Cloud, and call the Node-RED service you have already created from your dashboard.

2. Either click on **View all** or **Cloud Foundry services** on the **Resource summary** tile on the dashboard. Clicking on either option will take you to a list of resources on the IBM Cloud you created:

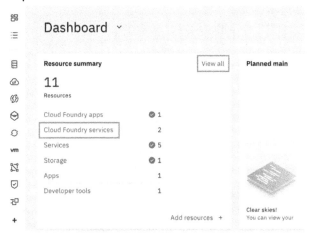

Figure 11.15 – Opening the resource list

If you have not created a Node-RED service on your IBM Cloud, please refer to *Chapter 6, Implementing Node-RED in the Cloud*, to create one before moving ahead.

3. Under the **Cloud Foundry apps** displayed on the **Resource list** screen, click on the Node-RED service you created to open the Node-RED flow editor:

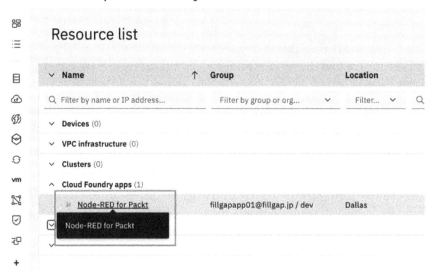

Figure 11.16 – Selecting the Node-RED service you created

4. Then, click **Visit App URL** to access the Node-RED flow editor:

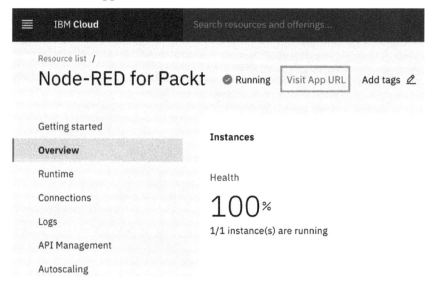

Figure 11.17 – Clicking Visit App URL

5. When the top screen of the Node-RED flow editor is displayed, click the **Go to your Node-RED flow editor** button to open the Node-RED flow editor:

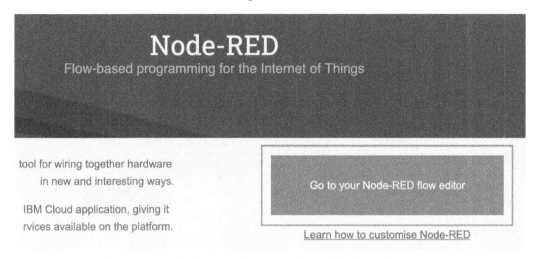

Figure 11.18 – Clicking the Go to your Node-RED flow editor button

6. Make a flow to visualize the data.

    When you accessed your Node-RED flow editor on your IBM Cloud, create a flow as follows. Place the **mqtt in** node, **json** node, two **change** nodes, and **gauge** node after each **change** node. If you want to get the debug log for this flow, please add the **debug** node after any node. In this example, two **debug** nodes are placed after the **mqtt in** node and the first **change** node.

    You already have the **dashboard** nodes, including the **gauge** node. If you don't have them, please go back to the *Make a flow for use case 2 – visualizing data* tutorial in *Chapter 6, Implementing Node-RED in the Cloud*, to get the **dashboard** nodes:

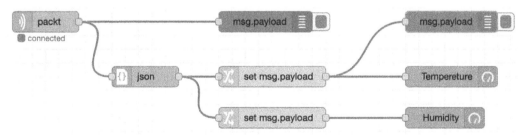

Figure 11.19 – Making a flow

7. Edit the **mqtt in** node. Double-click on the **mqtt in** node to open the settings window. Set **Topic**, **Qos**, and **Output** with the following values:

- **Topic**: `packt`

- **Qos**: `1`

- **Output**: `auto-detect` (string or buffer)

8. Click the **Edit** button (pencil icon) to the right of **Server** to open the credential properties:

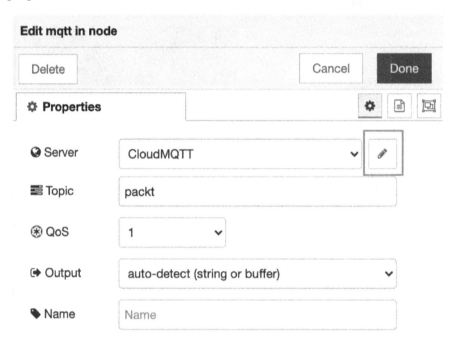

Figure 11.20 – Clicking the Edit button to open the Properties settings

9. On the Server settings panel, select the **Connection** tab, and fill in each property with the following values:

- **Server**: `driver.cloudmqtt.com`

- **Port**: `18913`

    The other properties of the **Connection** tab are not supposed to be changed and must be kept at their default values.

    You can refer to the following screenshot for the **Connection** tab settings:

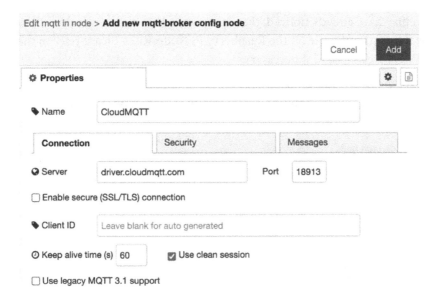

Figure 11.21 – MQTT broker server settings

10. Next, select the **Security** tab to edit the configuration to connect the MQTT server and fill in each property with the following values:

- **Username**: The user that you got from CloudMQTT.

- **Password**: The password that you got from CloudMQTT.

    You can refer to the following screenshot for the **Security** tab settings:

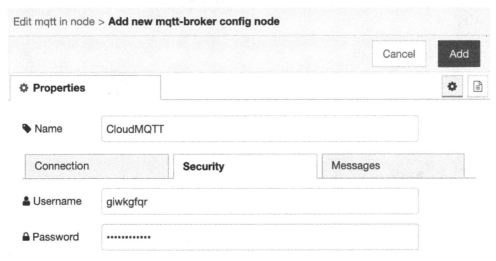

Figure 11.22 – MQTT broker user and password settings

As you may have already noticed, these properties have the same values that you set for the **mqtt out** node on the Raspberry Pi Node-RED. Please refer to the CloudMQTT dashboard if necessary.

11. Now, edit the json node. Double-click on the **json** node to open the settings window. Select **Convert between JSON String & Object** on **Action**, and set `msg.payload` on **Property**:

Figure 11.23 – Setting the json node properties

12. Edit the settings of the **change** node. Double-click on the first **change** node to open the **Settings** window and then set `msg.payload.temperature` in the box entitled **to** under the **Rules** area. Then, click the **Done** button to close the settings window:

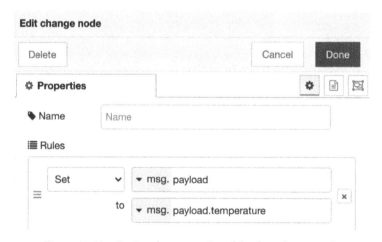

Figure 11.24 – Setting the properties of the first change node

13. Also, edit the settings of the second **change** node. Double-click on the second **change** node to open the **Settings** window. Set `msg.payload.humidity` in the box entitled **to** in the **Rules** area and then click the **Done** button to close the settings window:

Figure 11.25 – Setting the properties of the second change node

14. Edit the settings of the first **gauge** node. Double-click on the first **gauge** node to open the **Settings** window and then click the **Edit** button (pencil icon) to the right of **Group** to open the properties:

Figure 11.26 – Clicking the Edit button to open the Properties settings

15. In the dashboard's group setting panel, fill in each property with the following values:

- **Name:** `Raspberry Pi Sensor data`

  * It's OK to provide any name here. This name will be displayed on the chart web page that we'll create.

  Other properties are not supposed to be changed and must be kept at their default values. You can refer to the following screenshot:

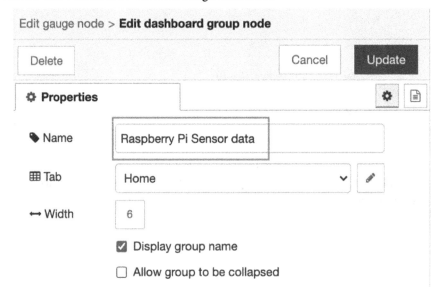

Figure 11.27 – Setting the group name

16. Go back to the main panel of the **gauge** node settings and fill in each property with the following values:

- **Type: Gauge**
- **Label:** `Temperature`
- **Units:** °C (if you prefer to use Fahrenheit, please use °F)
- **Range:** **-15 ~ 50** (if you prefer to use Fahrenheit, please adjust the range accordingly)

  Other properties are not supposed to be changed from their default values. You can refer to the following screenshot for the settings:

Figure 11.28 – Setting the gauge node properties

17. Edit the settings of the second **gauge** node. Double-click on the second **gauge** node to open the **Settings** window and then select the same **Group** name you created in the previous step. Fill in each property with the following values:

- **Type: Gauge**
- **Label**: Humidity
- **Units**: %
- **Range: 0 ~ 100**

Other properties are not supposed to be changed from their default values. You can refer to the following screenshot for the settings:

Figure 11.29 – Setting the gauge node properties

Please make sure to deploy the flow on your Node-RED.

This completes the Node-RED configuration on the IBM Cloud. This means that this flow is already subscribing (awaiting the data) with the topic packt for the CloudMQTT service. Next, it's time to publish and subscribe to the data.

Visualization of the data on the IBM Cloud

On the edge device side, on the Raspberry Pi, we are ready to publish the sensor data to CloudMQTT with the topic packt. On the cloud side, the flow is already with the packt topic for the CloudMQTT service.

For a Raspberry Pi, perform the following steps to publish your data:

1. Publish the data from your Raspberry Pi.

    Access your Node-RED flow editor on your Raspberry Pi. Click the button of the **inject** node to run this flow for publishing grove temperature and humidity sensor data:

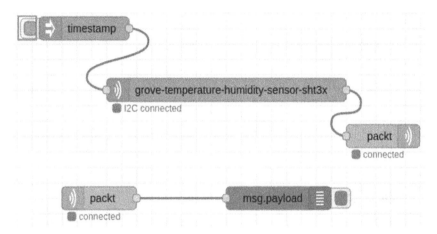

Figure 11.30 – Running the flow for publishing data

2. Check receipt of the data on the IBM Cloud.

    You will be able to receive (subscribe) the data via CloudMQTT. You can check it on the **debug** tab on your Node-RED flow editor on the IBM Cloud:

Figure 11.31 – Checking the subscribing of the data

3. Open the chart web page via the **chart** tab on your Node-RED flow editor on the IBM Cloud and then click the **open** button (diagonal arrow icon) to open it:

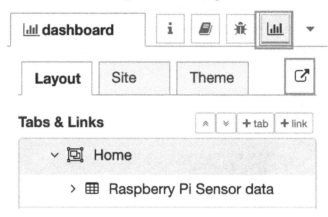

Figure 11.32 – Opening the chart web page

You will see the web page gauge chart with the data displayed:

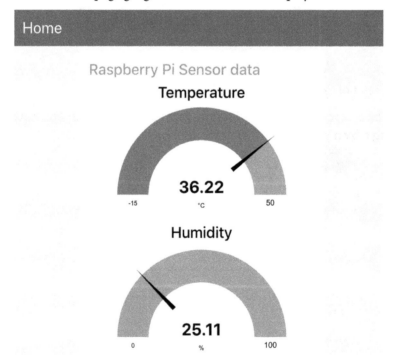

Figure 11.33 – The chart web page is displayed

Congratulations! Now you know how to observe the data sent from the Raspberry Pi on the server and visualize it as a chart.

If you want the flow configuration file to make this flow on your Node-RED, you can get it here: `https://github.com/PacktPublishing/-Practical-Node-RED-Programming/blob/master/Chapter11/getting-sensordata-with-iotplatform.json`.

Summary

In this chapter, we experienced how to receive the sensor data sent from an edge device and visualize it on the server side.

In this chapter, we used CloudMQTT and Node-RED on the IBM Cloud. Node-RED can run on any cloud platform and on-premises, and you can try to make this kind of application in any environment. That's why remembering this pattern will definitely be useful for your future development with other cloud IoT platforms.

In the next chapter, we will look at a hands-on scenario of making a chatbot application with Node-RED. This will introduce you to a new way of using Node-RED.

12
Developing a Chatbot Application Using Slack and IBM Watson

In this chapter, we will create a chatbot application, using Node-RED. For the chatbot application UI, we'll use Slack, and we'll use IBM Watson AI for skills. After completing the tutorials in this chapter, you will learn how to combine Node-RED with an external API to create an application. This will help you create extensible web applications with Node-RED in the future.

Let's get started with the following topics:

- Creating a Slack workspace
- Creating a Watson Assistant API
- Enabling a connection to Slack from Node-RED
- Building a chatbot application

By the end of this chapter, you will have mastered how to make a Slack chatbot application with Node-RED.

Technical requirements

To progress in this chapter, you will need the following:

- An IBM Cloud account: `https://cloud.ibm.com/`.
- The code used in this chapter can be found in `Chapter12` folder at `https://github.com/PacktPublishing/-Practical-Node-RED-Programming`.

Creating a Slack workspace

This hands-on tutorial uses **Slack** as the UI for your chatbot application. Node-RED is responsible for controlling the exchange of messages in the background of the chatbot application.

The overall view of this chatbot application is as follows:

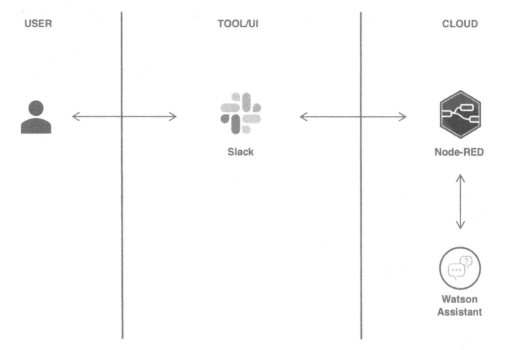

Figure 12.1 – Application overview

First of all, create a Slack workspace for use in this application with the following steps. If you already have a Slack workspace, you can use your existing one. In that case, skip the following steps and create a channel called `learning-node-red` in your workspace:

1. Access `https://slack.com/create`, enter your email address, and click the **Next** button:

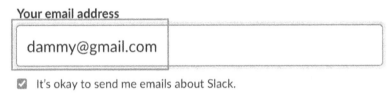

Figure 12.2 – Enter your email address

2. Check the six-digit verification code from the email you received from Slack:

Confirm your email address

Your confirmation code is below — enter it in the browser window where you've started signing up for Slack.

Questions about setting up Slack? Email us at help@slack.com

If you didn't request this email, there's nothing to worry about — you can safely ignore it.

Figure 12.3 – Check the six-digit code

3. Enter the verification code in the window that is displayed after you click **Next** with your email address. After entering your verification code, you will be redirected to the next window automatically:

Figure 12.4 – Enter the verification code

4. Give your workspace a name and click the **Next** button:

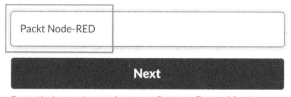

Figure 12.5 – Give your workspace a name

5. Create a channel in your workspace. You can use the **general** channel as it is, but let's create a channel to implement the chatbot. Here, we will create a channel named `Learning Node-RED`:

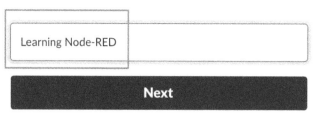

Figure 12.6 – Your workspace name

6. Click **skip for now** without adding teammates:

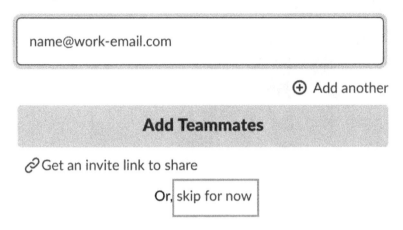

Figure 12.7 – No teammates are needed for this tutorial

7. Click **See Your Channel in Slack** to open the workspace you created:

Tada! Meet your team's first channel: #learning-node-red

You're leaving those unending email threads in the past. Channels give every project, topic, and team a dedicated space for all their messages and files.

Figure 12.8 – Click See Your Channel in Slack

You have created the workspace for this tutorial:

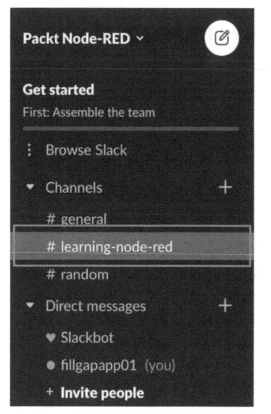

Figure 12.9 – You have created the workspace

> **Important note**
> The channel on which the chatbot resides should preferably be a channel that only you participate in unless you have a public purpose. This is because the chatbot's movement can be noisy for participants who do not like (or are not interested in) the chatbot.

At this point, you've got your workspace and channels ready to run your chatbot in Slack. Next, we will create a mechanism that will be the engine of the chatbot.

Creating a Watson Assistant API

This hands-on tutorial uses IBM's **Watson Assistant API** as the engine for chatbots. Watson Assistant can use natural language analysis to interpret the intent and purpose of natural conversation and return an appropriate answer.

Details about Watson Assistant can be found at the following URL: `https://www.ibm.com/cloud/watson-assistant-2/`.

To use the Watson Assistant API, you need to create an instance of the Watson Assistant API on IBM Cloud. Follow these steps to create it:

1. Log in to your IBM Cloud dashboard, and search `Assistant` in the **Catalog**. Click the **Assistant** tile on the results of your search:

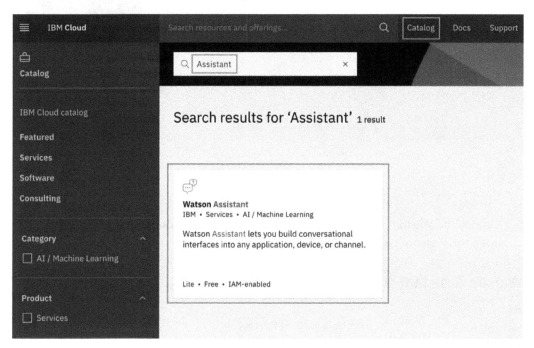

Figure 12.10 – Search for Watson Assistant

2. Create a Watson Assistant API service. Select a **region** for the Watson Assistant service data center. Dallas is stable so here we selected **Dallas**.

3. Select **Lite** for the pricing plan. Other items such as service name and resource group can be left at their default values.

4.   Click the **Create** button:

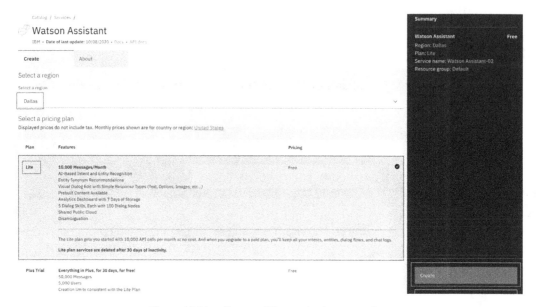

Figure 12.11 – Create a Watson Assistant service

5.   Launch the Watson Assistant tool. Click the **Launch Watson Assistant** button to open the Watson Assistant console:

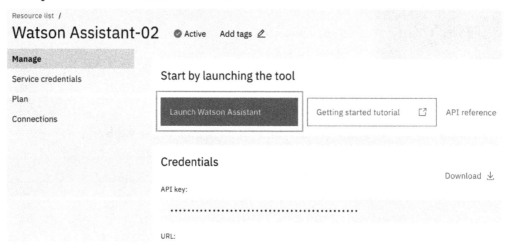

Figure 12.12 – Launch the Watson Assistant console

6.   Create a **Skill** in your **Watson Assistant** service.

You will be moved to the **My first skill** screen automatically when you open the Watson Assistant console for the first time.

Normally, you would create a Watson Assistant skill here, but this hands-on tutorial will focus on Node-RED rather than how to use Watson Assistant. Therefore, a skill in Watson Assistant is created by importing the definition file prepared in advance.

If you want to create your own skill, that's fine. In that case, the official Watson Assistant documentation will help: `https://cloud.ibm.com/apidocs/assistant/assistant-v2`.

7. Click **Assistants** on the side menu of the Watson Assistant console, and click the **Create assistant** button:

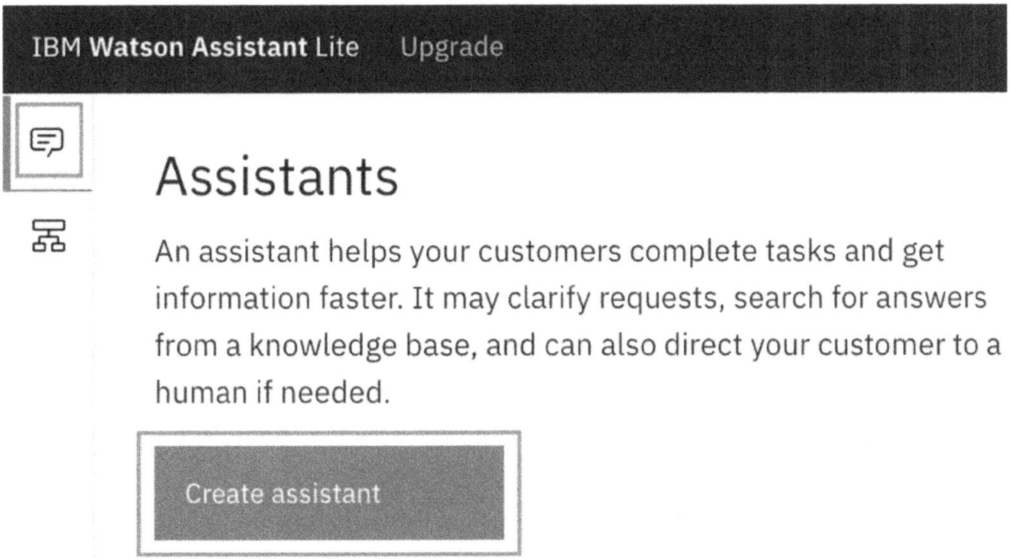

IBM **Watson Assistant** Lite   Upgrade

Assistants

An assistant helps your customers complete tasks and get information faster. It may clarify requests, search for answers from a knowledge base, and can also direct your customer to a human if needed.

Create assistant

Figure 12.13 – Create Assistant menu

This time, I prepared a skill that will randomly return a joke phrase when told `tell me a joke`.

8. Create an assistant for this frame, set the assistant's name to Respond Joke Phrase, and click the **Create assistant** button:

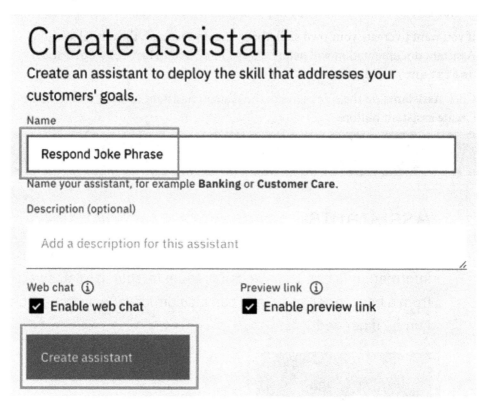

Figure 12.14 – Create assistant

9. Import **Dialog**. When your assistant is created, the settings screen of the created assistant is displayed. In the **Dialog** area on that settings screen, click the **Add dialog skill** button:

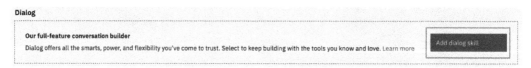

Figure 12.15 – Add dialog skill

10. Select the **Import skill** tab and select the JSON file for the skill you want to import. Download this JSON file at https://github.com/PacktPublishing/-Practical-Node-RED-Programming/blob/master/Chapter12/skill-Respond-Joke-Phrase.json.

11. Click the **Import** button when the JSON file is selected:

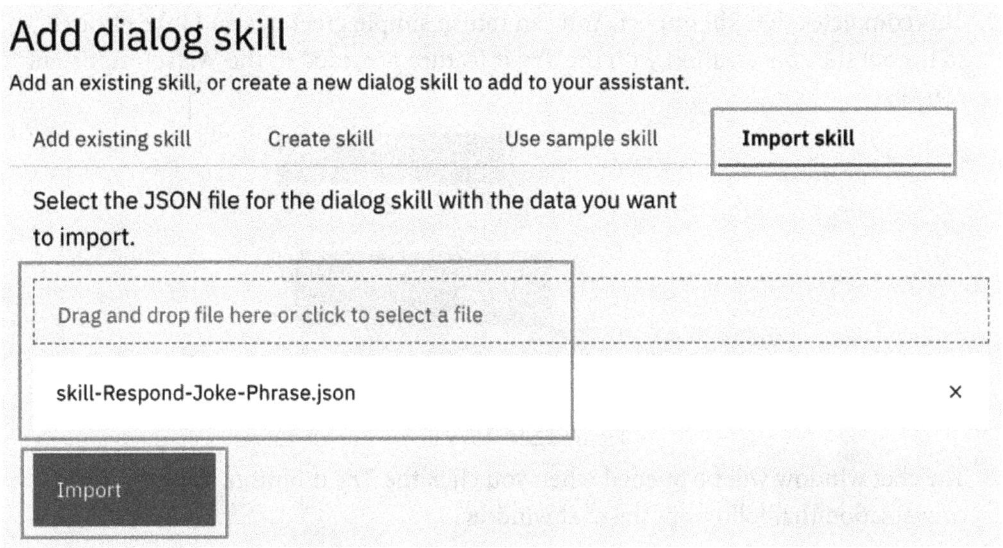

Add dialog skill

Add an existing skill, or create a new dialog skill to add to your assistant.

Add existing skill   Create skill   Use sample skill   **Import skill**

Select the JSON file for the dialog skill with the data you want to import.

Drag and drop file here or click to select a file

skill-Respond-Joke-Phrase.json   ×

Import

Figure 12.16 – Import the dialog skill file

You will see **Respond Joke Phrase** in the **Dialog** area:

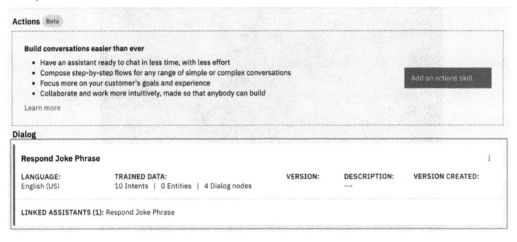

Figure 12.17 – Dialog skill imported

12. This completes the skill import. You can return simple greetings and joke phrases, so try out the conversation with the **Try it** feature provided in the Watson Assistant console:

Figure 12.18 – Try it

The chat window will be opened when you click the **Try it** button. Try typing the conversation that follows in the chat window:

`"Hello"; "Hi"; "Tell me jokes"; "Do you know any jokes?"`; *and so on…*

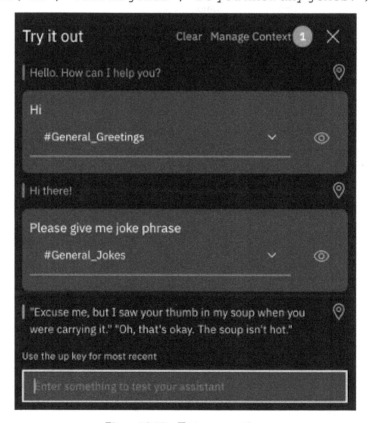

Figure 12.19 – Test conversation

If you don't get a good answer, try another phrase. Watson Natural Language Understanding divides conversations spoken in Watson Assistant's **Try it out** window into classes of intents or entities. If a conversation is not divided into the desired classes, you can train the Assistant API in the **Try it out** window.

Now that you've created an auto-answer conversation using Watson Assistant, there's one more thing to do, that is, confirmation of the Skill ID. This is the ID you will need later to operate Watson Assistant as an API from Node-RED.

Check the Skill ID from the **Skills** screen by following these steps:

1. Click **View API Details** under the **Skills** menu at the top right of the **Skill** tile you created:

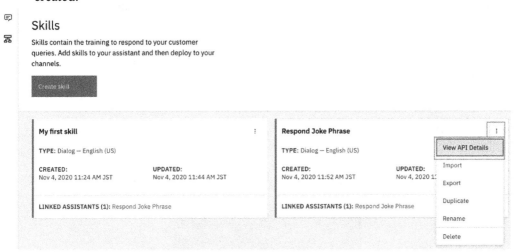

Figure 12.20 – Access the View API Details menu

2. Make a note of the **Skill ID** displayed:

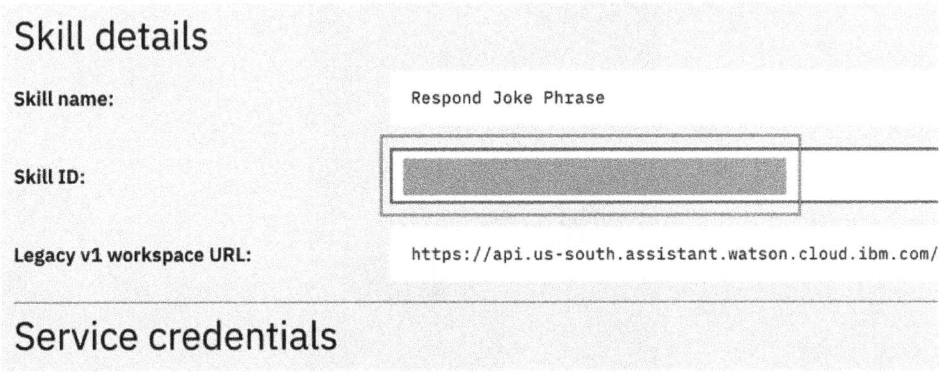

Figure 12.21 – Check and note the Skill ID

We have now created a chatbot service that automatically responds to chats. Next, let's integrate this with the Slack user interface.

Enabling the connection to Slack from Node-RED

Next, let's move on to the preparation of a Slack node on your Node-RED environment. Launch the Node-RED flow editor created on IBM Cloud.

What you do in this step is to install a node to connect to Slack in your Node-RED environment. The method is easy. All you have to do is find and install the node from the **Manage palette** window, which you've done several times in other chapters.

Follow these steps to proceed:

> **Important note**
>
> I believe that the Node-RED flow editor on your IBM Cloud has already been created as a service (as a Node.js application), but if you haven't done so already, refer to *Chapter 6, Implementing Node-RED in the Cloud*, to create a Node-RED service on IBM Cloud, before proceeding with this chapter.

1. You need to install the **node-red-contrib-slack** node to use Slack from Node-RED, so click **Manage palette**:

Figure 12.22 – Open the Manage palette window

2. Search the `node-red-contrib-slack` node and click the **Install** button:

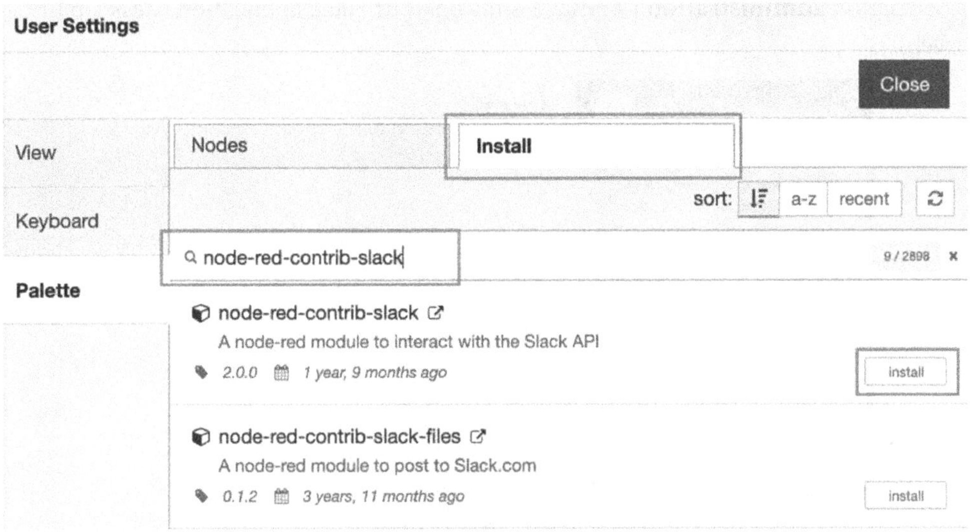

User Settings
Close
View
Keyboard
Palette
Nodes
Install
sort:
a-z
recent
node-red-contrib-slack
9 / 2898
node-red-contrib-slack
A node-red module to interact with the Slack API
2.0.0
1 year, 9 months ago
install
node-red-contrib-slack-files
A node-red module to post to Slack.com
0.1.2
3 years, 11 months ago
install

Figure 12.23 – Install the node-red-contrib-slack node

3. You will see four nodes that belong to **node-red-contrib-slack** on your palette. You have to prepare Slack nodes for building this sample application:

social
slack state
slack rtm in
slack rtm out
slack web out
twitter in
twitter out

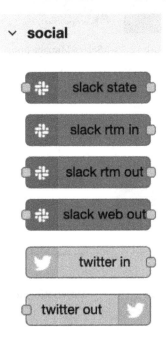

Figure 12.24 – Slack nodes will appear on your palette

4. Make a bot in your Slack workspace by accessing the **Slack App Directory** via **Settings & administration | Manage apps** on your Slack application (desktop or web):

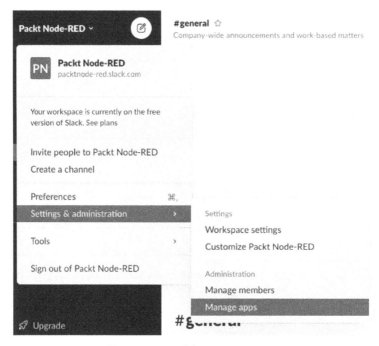

Figure 12.25 – Select Manage apps

5. After moving to the Slack App Directory website, click the **slack app directory** logo at the top left of the website to access the Slack App Directory main page:

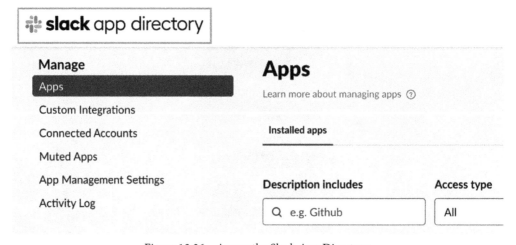

Figure 12.26 – Access the Slack App Directory

You can also access the Slack App Directory top page with the following URL: `https://<your workspace>.slack.com/apps`.

The following URL is just an example: `https://packtnode-red.slack.com/apps`.

This URL is generated automatically depending on each workspace name on Slack.

6. Click the **Get Essential Apps** button to move to the application search window:

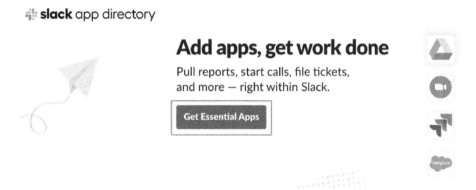

Figure 12.27 – Click the Get Essential Apps button

7. Search the word `bots` and click **Bots** on the results:

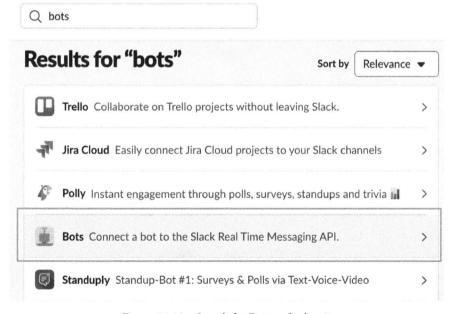

Figure 12.28 – Search for Bots and select it

8. Click the **Add to Slack** button on the **Bots** app screen:

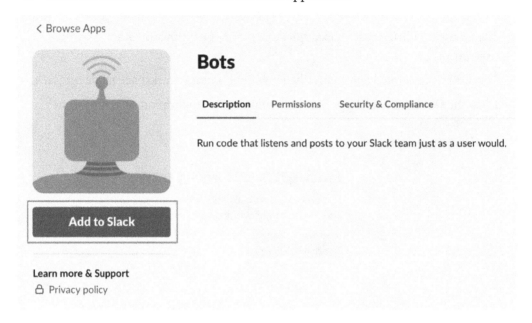

Figure 12.29 – Add the Bots app to your workspace

9. Set the **Username** of this bot application using any name you like. In this example, we named it `packt-bot`.

10. Click the **Add bot integration** button:

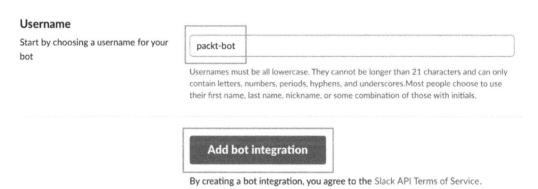

Figure 12.30 – Set your bot name

11. On the next screen, the API token for using the bot will be generated and displayed. Make a note of this so that you do not forget it. This API token is used when creating a flow with Node-RED:

> **Important note**
>
> Be careful when sharing bot user tokens with applications. Do not publish bot user tokens in public code repositories. This is because anyone can access the bot with this API token.

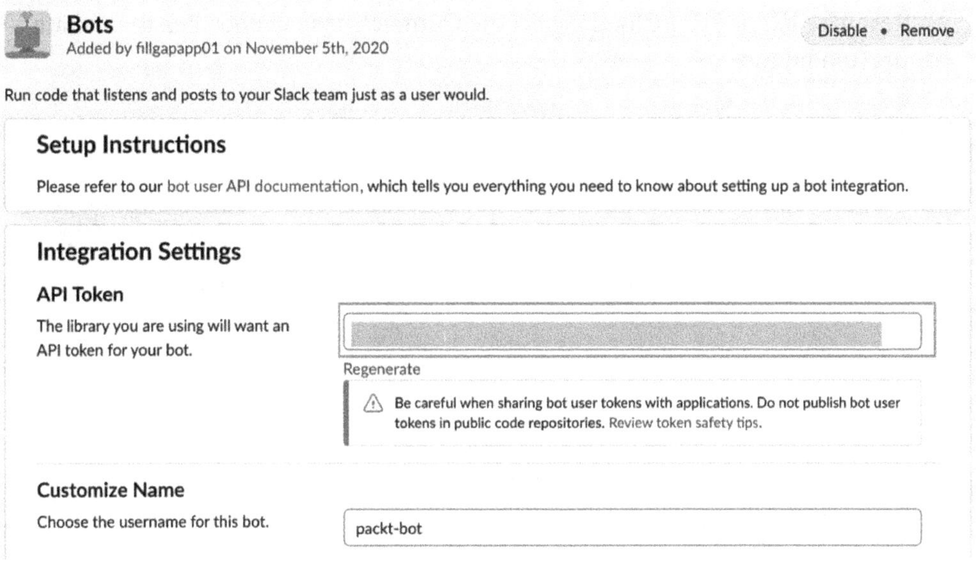

Bots   Disable • Remove
Added by fillgapapp01 on November 5th, 2020

Run code that listens and posts to your Slack team just as a user would.

Setup Instructions

Please refer to our bot user API documentation, which tells you everything you need to know about setting up a bot integration.

Integration Settings

API Token

The library you are using will want an API token for your bot.

Regenerate

⚠ Be careful when sharing bot user tokens with applications. Do not publish bot user tokens in public code repositories. Review token safety tips.

Customize Name

Choose the username for this bot.

packt-bot

Figure 12.31 – Confirm your API token

12. Click the **Save Integration** button to finish the bot app integration:

representing the same users.

Disable translation to send and receive global user IDs exclusively. Learn more.

Figure 12.32 – Bot app integration is finished

Now you are ready. Let's move on to the flow creation procedure.

Building a chatbot application

So far, you've created a chatbot engine in Watson Assistant, created a Slack workspace, and integrated the Bot app, which you can use in that Slack workspace.

Here, we will combine these services with Node-RED and create a mechanism with Node-RED so that the bot will answer when talking in Slack's workspace.

Follow these steps to create a flow:

1. Connect Watson Assistant to Node-RED. Access your Node-RED service dashboard via **Resource list** on IBM Cloud. Select the **Connections** tab and click the **Create connection** button:

Figure 12.33 – Create a new connection on Node-RED

2. Select the Watson Assistant service you created and click the **Next** button:

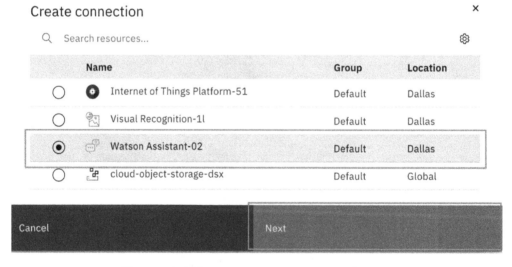

Figure 12.34 – Create a new connection on Node-RED

3. Click the **Connect** button with the default options to finish the connection setup. Doing this operation will restart the Node-RED application, which will take a few minutes to complete:

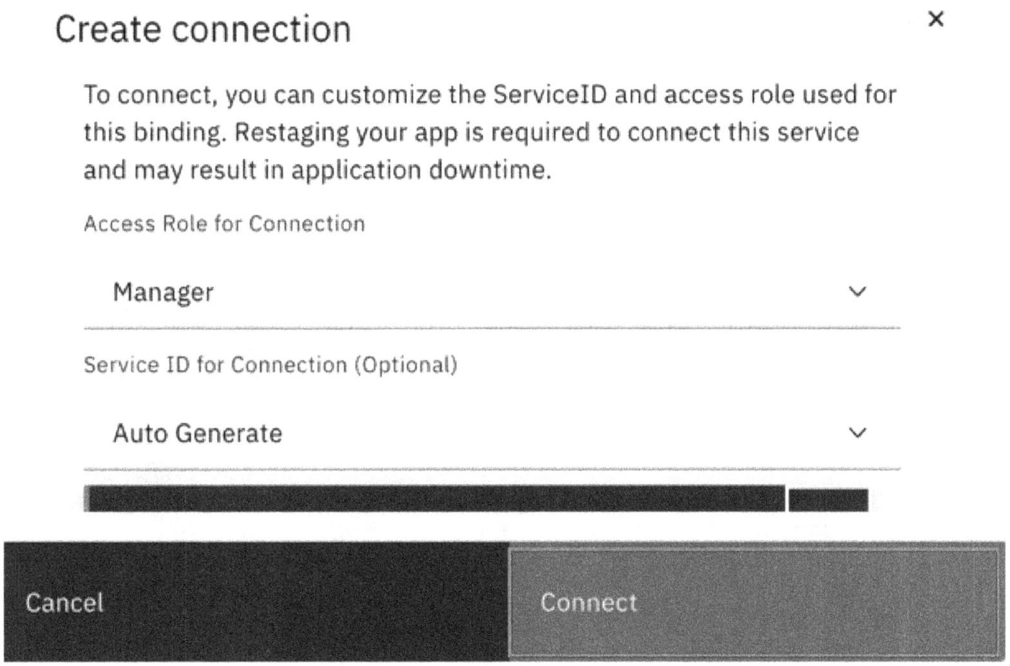

Create connection ✕

To connect, you can customize the ServiceID and access role used for this binding. Restaging your app is required to connect this service and may result in application downtime.

Access Role for Connection

Manager ⌄

Service ID for Connection (Optional)

Auto Generate ⌄

Cancel   Connect

Figure 12.35 – Finish creating the new connection on Node-RED

4. Make the flow to handle conversations on Slack.

    You already have Slack nodes and Watson nodes that are available to use for this hands-on tutorial.

5. Place a **slack-rtm-in** node, two **function** nodes, an **assistant** node, **slack-rtm-out**, and a **debug** node. After placing them, wire them sequentially as in the following figure:

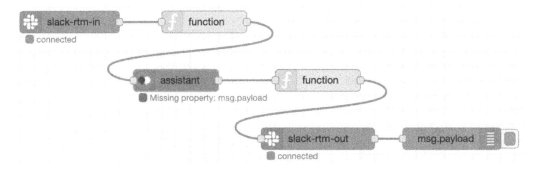

Figure 12.36 – Place the nodes and wire them

6. Set the parameters for each node.

Follow this procedure to set the parameters on each node. For the nodes that need to be coded, code them as follows:

- The **slack-rtm-in** node:

a) Click the edit button (pencil icon) to open the **Properties** panel:

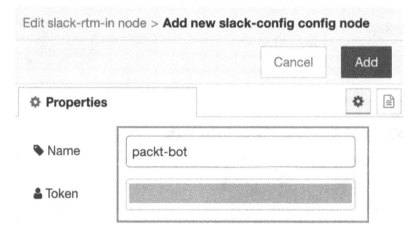

Figure 12.37 – Open the Properties panel

b) Enter the **Token** value you generated on your Slack Bots app. You can set any name for this configuration. In the example here, it's named `packt-bot`:

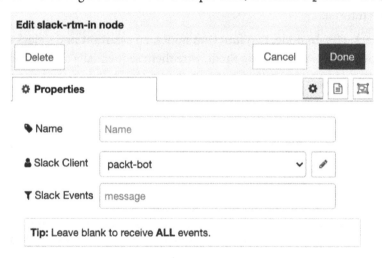

Figure 12.38 – Set the properties of the configuration to connect the Slack app

When you go back to the main panel of this node, you will see the configuration has been set in the **Slack Client** property.

c) Click the **Done** button to close this setting:

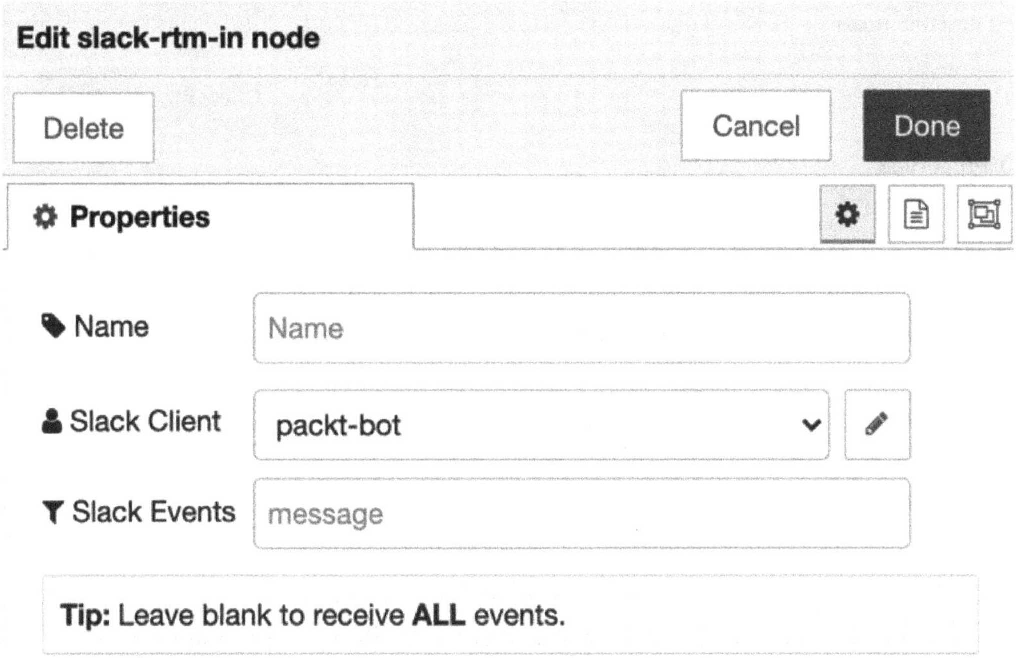

Edit slack-rtm-in node

| Delete | | Cancel | Done |

⚙ **Properties**   ⚙ 📄 🔲

🏷 Name | Name

👤 Slack Client | packt-bot   ∨ ✏

🔻 Slack Events | message

Tip: Leave blank to receive **ALL** events.

Figure 12.39 – Finish setting the properties of the slack-rtm-in node

- The **function** node (first one):

a) In the first **function** node, enter the following:

```
global.set("channel",msg.payload.channel);

msg.topic = "message";
msg.payload = msg.payload.text;
return msg
```

You can also refer to the following figure:

Figure 12.40 – First function node coding

In this function node, the message that is posted on Slack is taken out from the JSON data sent from Slack and put in `msg.payload` again.

Another important process is to store the channel information sent from Slack in the Global variable in Node-RED. The channel information stored here will be used later when sending a response message back to Slack.

- The **assistant** node:

In the previous step, you connected Watson Assistant to Node-RED. This means that you can call the Assistant API from Node-RED without using an API key or secret.

When I double-click the **assistant** node to open the settings panel, I don't see any properties such as API keys. If you see them in your settings panel, it means that the Watson Assistant and Node-RED connection process is failing. In that case, perform the connection process again.

There is only one property to set here. Set the Watson Assistant Skill ID you wrote down earlier as the **Workspace ID** property in the **assistant** node's settings panel:

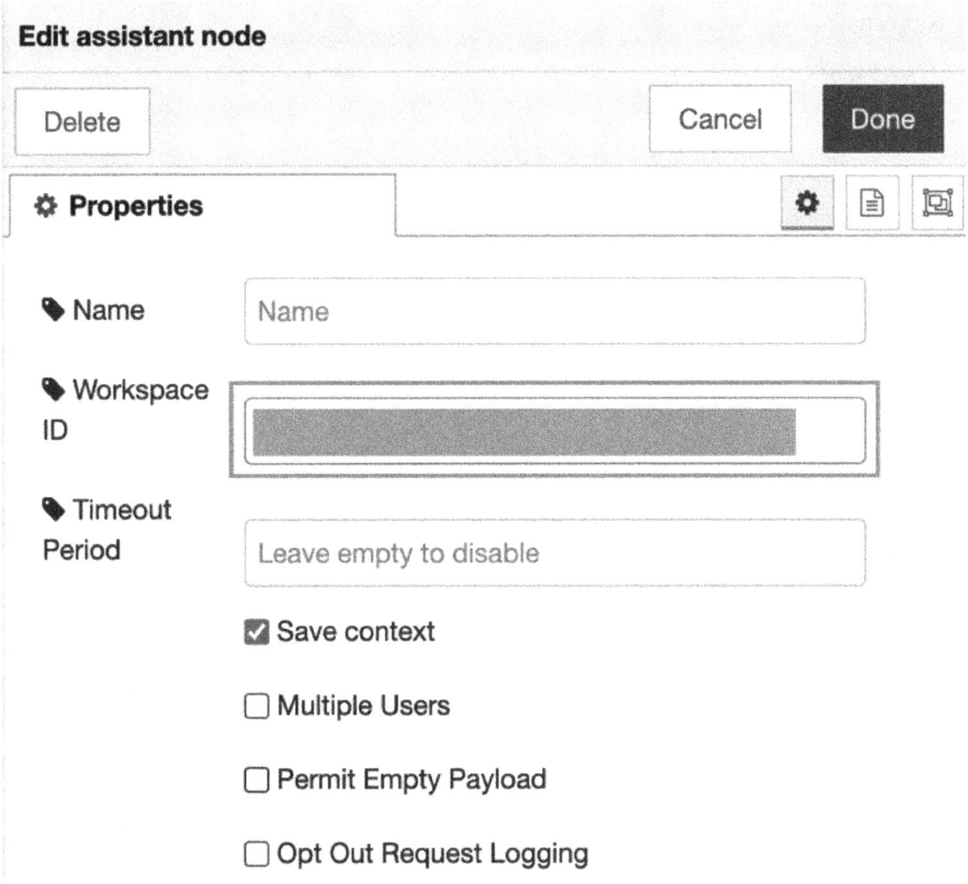

Edit assistant node

| Delete | | Cancel | Done |

⚙ **Properties**

🏷 Name

| Name |

🏷 Workspace ID

🏷 Timeout Period

| Leave empty to disable |

☑ Save context

☐ Multiple Users

☐ Permit Empty Payload

☐ Opt Out Request Logging

Figure 12.41 – Set the Skill ID as the Workspace ID

This completes the settings for the **assistant** node. Save your settings and close the settings panel.

- The **function** node (the second one):

In the first **function** node, enter the following code:

```
var g_channel=global.get("channel");

msg.topic = "message";
msg.payload = {
    channel: g_channel,
    text: msg.payload.output.text[0]
}
return msg
```

You can also refer to the following figure:

Edit function node

| Delete | | Cancel | Done |

☼ **Properties**

🏷 Name   [Name]

| Setup | Function | Close |

```
1  var g_channel=global.get("channel");
2
3  msg.topic = "message";
4  msg.payload = {
5      channel: g_channel,
6      text: msg.payload.output.text[0]
7  }
8  return msg
```

Figure 12.42 – Second function node coding

The second function node stores the autoresponder message returned from Watson Assistant in msg.payload.text, and gets the Slack channel information saved in the first function node and stores it in msg.payload.channel.

- The **slack-rtm-out** node:

Next is the **slack-rtm-out** node, which is easy to set up:

a) Double-click on the **slack-rtm-out** node to open the settings panel.

b) You will see that the configuration named packt-bot you created is already placed in this node property. If it is not set yet, please select it from the drop-down list manually. Once you click on **Done**, the settings will be complete:

Figure 12.43 – Check the property settings of the slack-rtm-out node

- The **debug** node:

  The debug node here simply outputs the log. No settings are required.

7. Check the bot application on Slack.

   An auto-answer chatbot has been created using Slack. Let's try the conversation.

8. On the channel you created in your Slack workspace, add the bot app you integrated and click the **Add an app** link on the channel:

#learning-node-red

You created this channel on October 30th. This is the very beginning

✎ Add description | ✄⁺ Add an app | ዲ₊ Add people

Figure 12.44 – Click the Add an app link

9. Click the **Add** button to add the bot app to your channel:

In your workspace (1)

packt-bot   **Add**

Figure 12.45 – Add the bot app you created

Now, let's actually have a conversation. Mention and talk to your bot (packt-bot in the example) on the channel where you added this bot app. Since the only conversations we are learning with Watson Assistant this time are greetings and listening to jokes, we will send a message from Slack that seems to be related to either of these.

First, let's say Hello. You will see a greeting kind of response:

fillgapapp01 11:49 AM
@packt-bot Hello

packt-bot APP 11:49 AM
Hello my friend!

Figure 12.46 – Exchanging greetings with the chatbot

Then send a message like Please tell me a joke. It randomly responds with a bot-selected joke as a reply:

 fillgapapp01 11:50 AM
Please tell me a joke

 packt-bot APP 11:50 AM
"Why did the cucumber blush?" "Because it saw the salad dressing."

Figure 12.47 – The chatbot answers some jokes

Great work! You finally made the chatbot application with Node-RED.

If you want the flow configuration file to make this flow in your Node-RED environment, you can get it here: https://github.com/PacktPublishing/-Practical-Node-RED-Programming/blob/master/Chapter12/slack-watson-chatbot-flows.json.

Summary

In this chapter, we experienced how to make a chatbot application with Slack, Watson, and Node-RED. This time, we used Slack as a chat platform, but you can use any chat platforms that have APIs, such as LINE, Microsoft Teams, and so on, instead of Slack.

This chapter is also very helpful for creating any applications that are not IoT-based. Node-RED can develop various applications by linking with any Web API.

In the next chapter, let's develop our own node. Of course, it can be used in any environment. Developing your own node with Node-RED means developing a new node that cannot be done with the existing nodes. This is surely the first step for advanced users of Node-RED.

13
Creating and Publishing Your Own Node on the Node-RED Library

So far, we have learned about Node-RED using the prepared nodes. In this chapter, you'll learn how to create your own node and publish it in a library. After completing the tutorials in this chapter, you will be able to publish your own node for use by various developers around the world.

Let's get started with the following topics:

- Creating your own node
- Testing your own node in a local environment
- Publishing your own node as a module on the Node-RED Library

By the end of this chapter, you will have mastered how to create your own node.

Technical requirements

To progress in this chapter, you will need the following:

- A GitHub account: `https://github.com/`.

- An npm account: `https://www.npmjs.com/`.

- Node-RED (standalone in a local environment).

- An IBM Cloud account.

- The code used in this chapter can be found in `Chapter13` folder at `https://github.com/PacktPublishing/-Practical-Node-RED-Programming`.

- The steps of this tutorial are basically processed on Mac. If you use a Windows PC, please replace the commands and file path with your environment.

Creating your own node

Before developing a node, there is something you need to know first. The following policies are set for node development. Let's follow these and develop a node.

When creating a new node, you need to follow some general rules. They adhere to the approach adopted by the core nodes and provide a consistent user experience.

You can check the rules for creating a node on the official Node-RED website: `https://nodered.org/docs/creating-nodes/`.

Node program development

Node-RED nodes consist of two files: a JavaScript file that defines processing and an HTML file that provides a UI such as a setting screen. In the JavaScript file, the processing of the node you create is responsible for is defined as a function. This function is passed an object that contains node-specific properties. The HTML file describes the property settings screen displayed by the Node-RED flow editor. The settings values entered on the property settings screen displayed in this HTML file are called from the JavaScript file and processed.

Here, we will create a GitHub repository, but if you just want to create a node, you don't need a GitHub repository. In this chapter, we will use the GitHub repository to publish the created node to the library, so I would like you to create the repository at the beginning of the step.

Please implement the following steps to create a GitHub repository:

1.   Go to `https://github.com/` and log in with your GitHub account.

2.   Select **New repository** from the + dropdown at the top right of the GitHub page:

Figure 13.1 – Create a repository for your own node

The repository created here exists as a project for developing nodes, and then it will be packaged and published to npm. (Of course, it is optional to publish it.)

Therefore, make sure that the repository name follows the naming convention for node development.

The GitHub repository name will be the same as the node name. In the node creation rule, the node name must be `node-red-contrib-<name representing a group of nodes>`, so specify the GitHub repository name accordingly. In this example, it is `node-red-contrib-taiponrock`.

3.   After specifying the repository name, set the repository disclosure range to **Public**, check the README file, and specify the license. In this example, it is created with the **Apache License 2.0**.

4. After setting everything, click the **Create repository** button to create a repository:

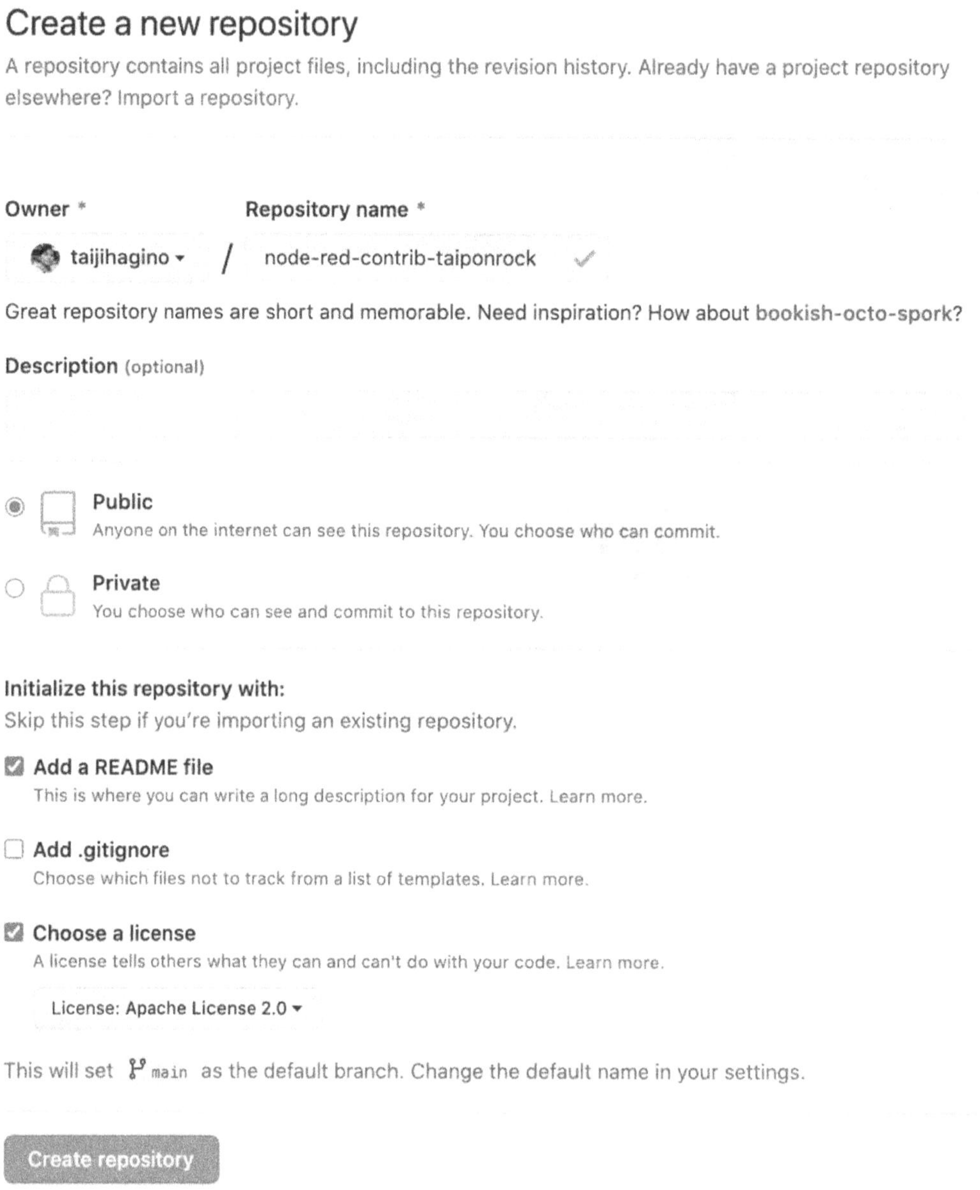

Create a new repository

A repository contains all project files, including the revision history. Already have a project repository elsewhere? Import a repository.

Owner *   Repository name *

🔵 taijihagino ▾   /   node-red-contrib-taiponrock   ✓

Great repository names are short and memorable. Need inspiration? How about bookish-octo-spork?

Description (optional)

◉ 📖 **Public**
      Anyone on the internet can see this repository. You choose who can commit.

○ 🔒 **Private**
      You choose who can see and commit to this repository.

Initialize this repository with:
Skip this step if you're importing an existing repository.

☑ **Add a README file**
  This is where you can write a long description for your project. Learn more.

☐ **Add .gitignore**
  Choose which files not to track from a list of templates. Learn more.

☑ **Choose a license**
  A license tells others what they can and can't do with your code. Learn more.

  License: Apache License 2.0 ▾

This will set ⑂ main as the default branch. Change the default name in your settings.

[Create repository]

Figure 13.2 – The repository is created as a public project

You have now created your GitHub repository.

Now let's clone the repository we just created to our local development environment by following these steps:

1. Copy the repository URL to the clipboard. Click the green **Code** dropdown and click the **clipboard** button to copy the URL:

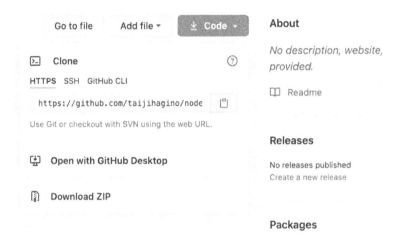

Figure 13.3 – Copy the URL to clone this repository

Clone the repository locally (git clone) from a command-line interface (such as a terminal) where **Bash** can run.

2. Go to the working directory where you want to clone (copy) the repository. Here, I created a work directory under the user directory and moved to it:

```
$ mkdir work
$ cd work
```

3. Execute the git clone command with the URL of the repository you created earlier:

```
$ git clone https://github.com/<GitHub account>/node-red-
contrib-<Any specified string>.git
```

4. When the clone is finished, use the ls command to confirm that it has been successfully cloned:

```
$ls
node-red-contrib-<Any specified string>
```

Let's make a JavaScript file now.

From here, we will create the actual node processing. But don't worry, we already have the code ready. The provided code is very simple for processing the node. It's just a matter of converting the string passed as input to lowercase.

5. First, change to the directory of the cloned repository:

```
$ cd node-red-contrib-<arbitrary specified string>
```

6. Under this directory, create a file with the filename node.js, as shown in the following code:

```
module.exports = function(RED) {
    function LowerCaseNode(config) {
        RED.nodes.createNode(this,config);
        var node = this;
        node.on('input', function(msg) {
            msg.payload = msg.payload.toLowerCase();
            node.send(msg);
        });
    }
    RED.nodes.registerType("lower-case",LowerCaseNode);
}
```

node.js has been created.

Let's make an HTML file now.

7. Create a file under the same directory with the filename node.html, as shown in the following code:

```
<script type="text/javascript">
    RED.nodes.registerType('lower-case',{
        category: 'function',
        color: '#a6bbcf',
        defaults: {
            name: {value:""}
        },
        inputs:1,
        outputs:1,
        icon: "file.png",
        label: function() {
```

```
                    return this.name||"lower-case";
        }
    });
</script>


<script type="text/html" data-template-name="lower-case">
    <div class="form-row">
        <label for="node-input-name"><i class="icon-
            tag"></i> Name</label>
        <input type="text" id="node-input-name"
            placeholder="Name">
    </div>
</script>


<script type="text/html" data-help-name="lower-case">
    <p>A simple node that converts the message payloads
        into all lower-case characters</p>
</script>
```

node.html has been created. This HTML file is responsible for the UI and design of the node you create. As mentioned previously, a node always consists of an HTML file and a JavaScript file.

The node implementation has been almost completed. Next, let's package the created node so that it can be deployed.

Node packaging

Now that we've created the node processing (JavaScript) and appearance (HTML), it's time to package it. In Node-RED, the flow editor itself is a **Node.js** app, and each node running on it is also a Node.js app. In other words, the packaging here is processed using npm.

We won't go into detail about npm here. If you want to know more about it, please visit the npm official website or refer to various technical articles: https://www.npmjs.com/.

Now, use the npm command to perform the following steps:

1. npm initialization. Execute the following command in the same location as the directory where node.js and node.html were created:

    ```
    $ npm init
    ```

2. When you run npm init, you will be asked for various parameters interactively, so enter them according to how you want to proceed. These are the parameters that I used:

Parameter	Value
name	Default
version	Default
description	Node description (displayed as a library or installation description)
entry point	Default
test command	No input required
git repository	Default
keywords	Specify keywords used for search during installation, separated by commas. This time, we are planning to register the library, so be sure to enter node-red.
author	npm account
license	Apache-2.0 (Default)

When you finish this step, the npm init command will generate a package.json file:

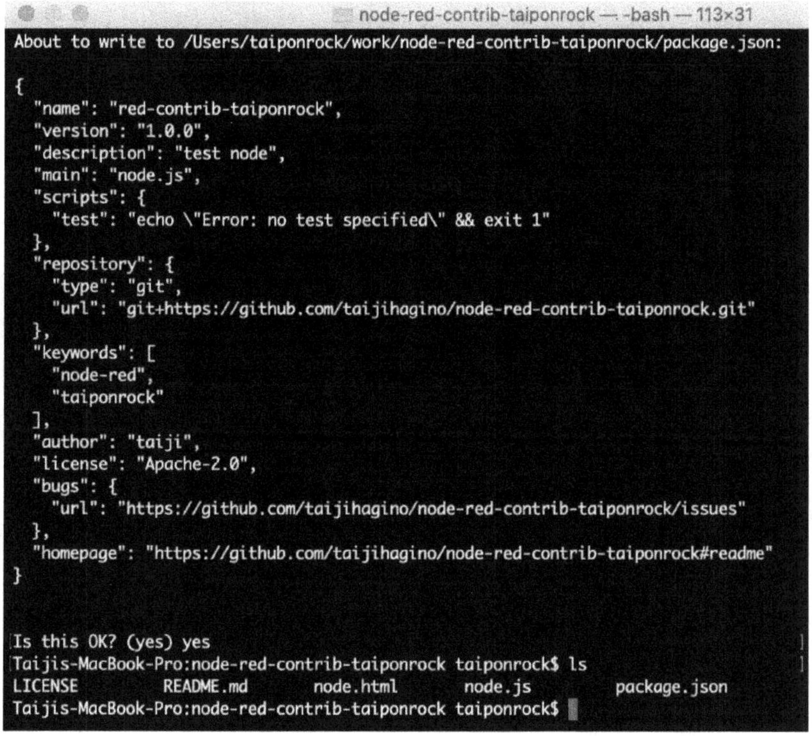
node-red-contrib-taiponrock — -bash — 113×31
About to write to /Users/taiponrock/work/node-red-contrib-taiponrock/package.json:
{
"name": "red-contrib-taiponrock",
"version": "1.0.0",
"description": "test node",
"main": "node.js",
"scripts": {
"test": "echo \"Error: no test specified\" && exit 1"
},
"repository": {
"type": "git",
"url": "git+https://github.com/taijihagino/node-red-contrib-taiponrock.git"
},
"keywords": [
"node-red",
"taiponrock"
],
"author": "taiji",
"license": "Apache-2.0",
"bugs": {
"url": "https://github.com/taijihagino/node-red-contrib-taiponrock/issues"
},
"homepage": "https://github.com/taijihagino/node-red-contrib-taiponrock#readme"
}
Is this OK? (yes) yes
Taijis-MacBook-Pro:node-red-contrib-taiponrock taiponrock$ ls
LICENSE README.md node.html node.js package.json
Taijis-MacBook-Pro:node-red-contrib-taiponrock taiponrock$

Figure 13.4 – npm init

3. Edit `package.json`. You will need to manually add Node-RED-specific settings to `package.json`. Open the `package.json` file with a text editor and add the new property at the same level as `"name"` and `"version"` in the JSON: `"node-red": {"nodes": "{" lower-case ":" node.js "} }`:

```
{
    "name": "node-red-contrib-<arbitrary string
      specified>",
    "version": "1.0.0",
    (abridgement)
    "node-red": {
      "nodes": {
        "lower-case": "node.js"
      }
    },
    (abridgement)
}
```

The following screenshot can be used as a reference, which will help you in adding this property:

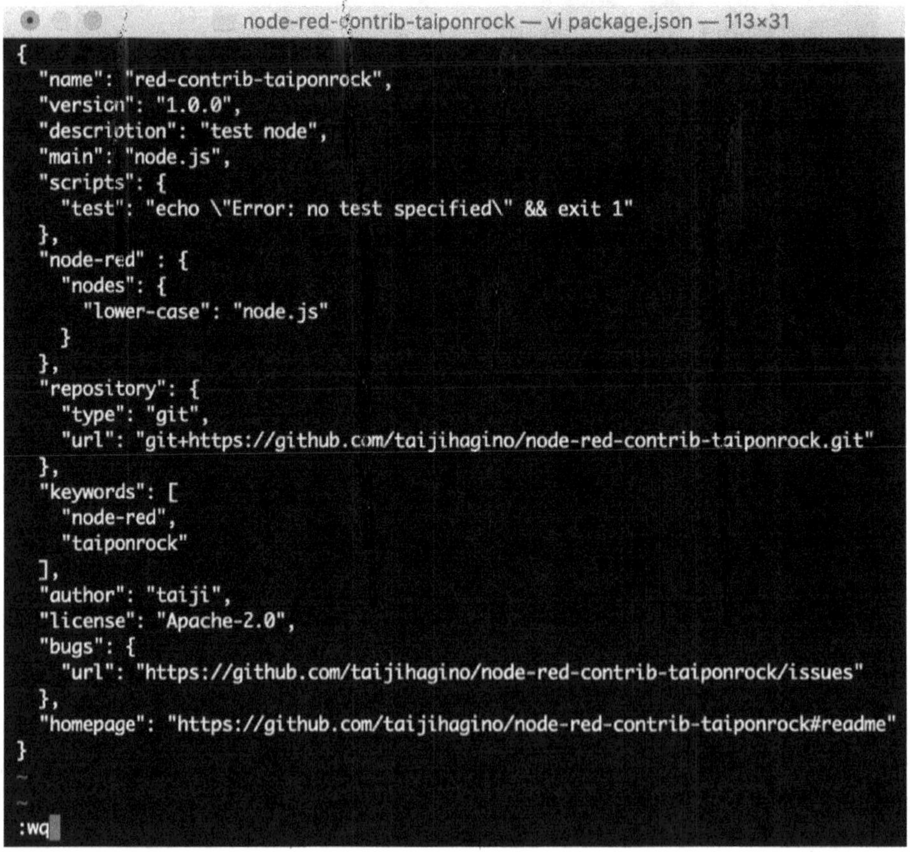

```
node-red-contrib-taiponrock — vi package.json — 113×31
{
  "name": "red-contrib-taiponrock",
  "version": "1.0.0",
  "description": "test node",
  "main": "node.js",
  "scripts": {
    "test": "echo \"Error: no test specified\" && exit 1"
  },
  "node-red" : {
    "nodes": {
      "lower-case": "node.js"
    }
  },
  "repository": {
    "type": "git",
    "url": "git+https://github.com/taijihagino/node-red-contrib-taiponrock.git"
  },
  "keywords": [
    "node-red",
    "taiponrock"
  ],
  "author": "taiji",
  "license": "Apache-2.0",
  "bugs": {
    "url": "https://github.com/taijihagino/node-red-contrib-taiponrock/issues"
  },
  "homepage": "https://github.com/taijihagino/node-red-contrib-taiponrock#readme"
}
~
:wq
```

Figure 13.5 – Edit package.json

This completes the packaging of your own node. Let's actually use this node in the next part.

Testing your own node in a local environment

You have already completed your own node. Let's add the nodes created so far to Node-RED in a local environment.

For your own nodes, it is very important to check their operation locally. Publishing a node on the internet without making sure it works in your environment is not good for many developers.

So, in this section, you'll be testing your own node in your local environment.

Node installation

You can use the npm link command to test the node module locally. This allows you to develop nodes in your local directory and link them to your local Node-RED installation during development.

This is very simple. Follow these steps:

1. Execute the following command on the CLI to add a node and start Node-RED:

```
$ cd <path to node module>
$ npm link
```

   This will create the appropriate symbolic link to the directory and Node-RED will discover the node at boot time. Simply restart Node-RED to get the changes to the node's files.

2. Run the node-red command on the command line to start the local Node-RED. If it has already been started, restart it.

   You should see that a node called **lower case** has been added to the **function** category of the palette after rebooting:

Figure 13.6 – The lower case node has been added

3. Let's see if it can be used properly. Create a flow by sequentially connecting each node of **inject lower case debug**.

4. For the properties of the **inject** node, set it to the character string type and set it to output any character string in all uppercase letters, for example, **MY NAME IS TAIJI**:

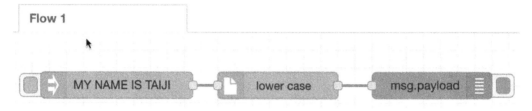

Figure 13.7 – Make a flow

5. When you deploy the created flow and execute the **inject** node, you can see that the all-uppercase string, as the parameter of this flow, is converted to an all-lowercase string and output to the **debug** tab:

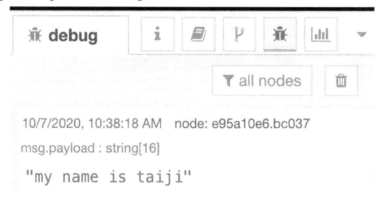

Figure 13.8 – Result of this flow

Next, let's see how to customize a node.

Node customization

I was able to confirm that the node I created can be used in the local environment. From here, we will customize that node. It is possible to edit the function and appearance of the node by modifying JavaScript and HTML. These changes will take effect when you restart Node-RED.

Changing the node name

Currently, the node name of the created node is still a lower-case version of the sample program. Here, change this name to any name you like. Every node must have a unique name, so you should pick something that does not already exist. Follow these steps to change the name of the node:

1. Change `lower-case` described in the `package.json` file to your own node name.

   In the example, the repository of the node is `node-red-contrib-taiponrock`, so change it to the `taiponrock` node.

   This is how the `package.json` file looks before being modified:

```
node-red-contrib-taiponrock — vi package.json — 113×31
{
  "name": "red-contrib-taiponrock",
  "version": "1.0.0",
  "description": "test node",
  "main": "node.js",
  "scripts": {
    "test": "echo \"Error: no test specified\" && exit 1"
  },
  "node-red": {
    "nodes": {
      "lower-case": "node.js"
    }
  },
  "repository": {
    "type": "git",
    "url": "git+https://github.com/taijihagino/node-red-contrib-taiponrock.git"
  },
  "keywords": [
    "node-red",
    "taiponrock"
  ],
  "author": "taiji",
  "license": "Apache-2.0",
  "bugs": {
    "url": "https://github.com/taijihagino/node-red-contrib-taiponrock/issues"
  },
  "homepage": "https://github.com/taijihagino/node-red-contrib-taiponrock#readme"
}
~
~
"package.json" 28L, 665C
```

Figure 13.9 – Before modifying package.json

And this is how it looks after being modified:

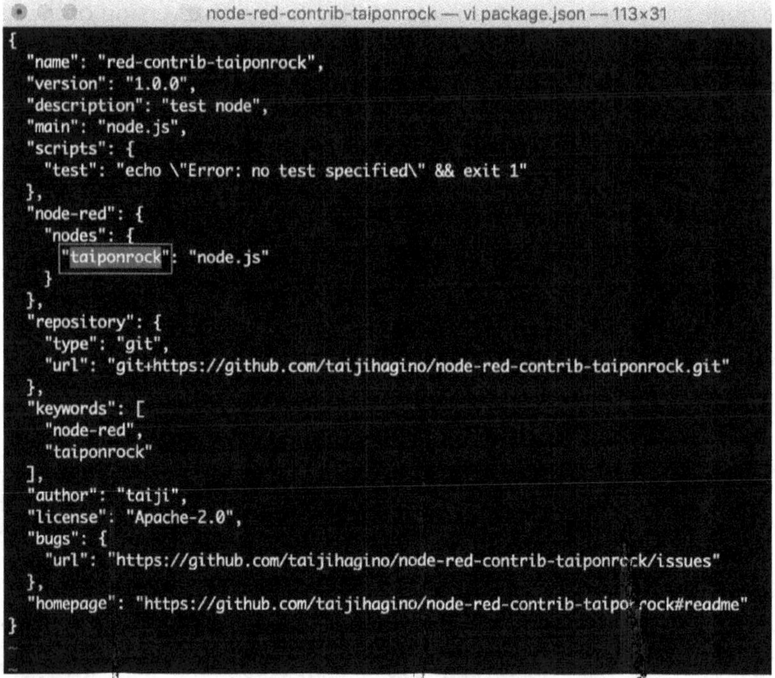

```
node-red-contrib-taiponrock — vi package.json — 113×31
{
  "name": "red-contrib-taiponrock",
  "version": "1.0.0",
  "description": "test node",
  "main": "node.js",
  "scripts": {
    "test": "echo \"Error: no test specified\" && exit 1"
  },
  "node-red": {
    "nodes": {
      "taiponrock": "node.js"
    }
  },
  "repository": {
    "type": "git",
    "url": "git+https://github.com/taijihagino/node-red-contrib-taiponrock.git"
  },
  "keywords": [
    "node-red",
    "taiponrock"
  ],
  "author": "taiji",
  "license": "Apache-2.0",
  "bugs": {
    "url": "https://github.com/taijihagino/node-red-contrib-taiponrock/issues"
  },
  "homepage": "https://github.com/taijihagino/node-red-contrib-taiponrock#readme"
}
```

Figure 13.10 – After modifying package.json

2. Change `lower-case` and `LowerCaseNode` described in the `node.js` file to your own node name.

    For example, change `lower-case` to `taiponrock` and `LowerCaseNode` to `TaiponrockNode`.

    This is how the `node.js` file looks before being modified:

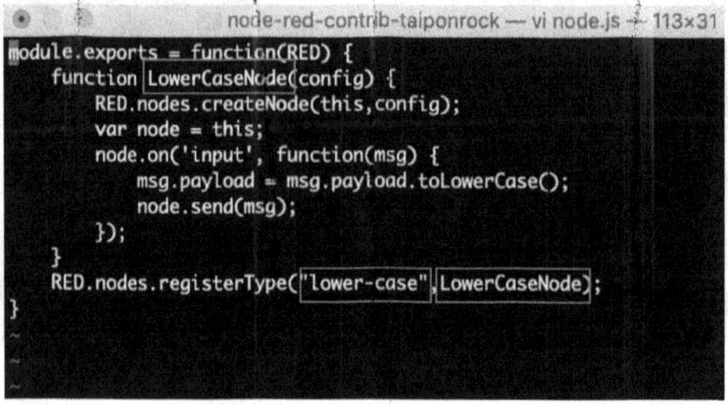

```
node-red-contrib-taiponrock — vi node.js — 113×31
module.exports = function(RED) {
    function LowerCaseNode(config) {
        RED.nodes.createNode(this,config);
        var node = this;
        node.on('input', function(msg) {
            msg.payload = msg.payload.toLowerCase();
            node.send(msg);
        });
    }
    RED.nodes.registerType("lower-case",LowerCaseNode);
}
```

Figure 13.11 – Before modifying node.js

This is how the `node.js` file looks like after being modified:

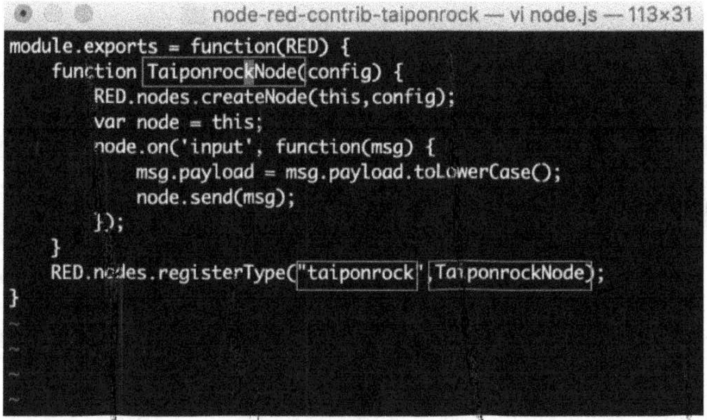
node-red-contrib-taiponrock — vi node.js — 113×31
module.exports = function(RED) {
function TaiponrockNode(config) {
RED.nodes.createNode(this,config);
var node = this;
node.on('input', function(msg) {
msg.payload = msg.payload.toLowerCase();
node.send(msg);
});
}
RED.nodes.registerType("taiponrock",TaiponrockNode);
}

Figure 13.12 – After modifying node.js

3. Change `lower-case` described in the `node.html` file to your own node name.

   For example, change `lower-case` to `taiponrock`.

   This is how the `node.html` file looks before being modified:

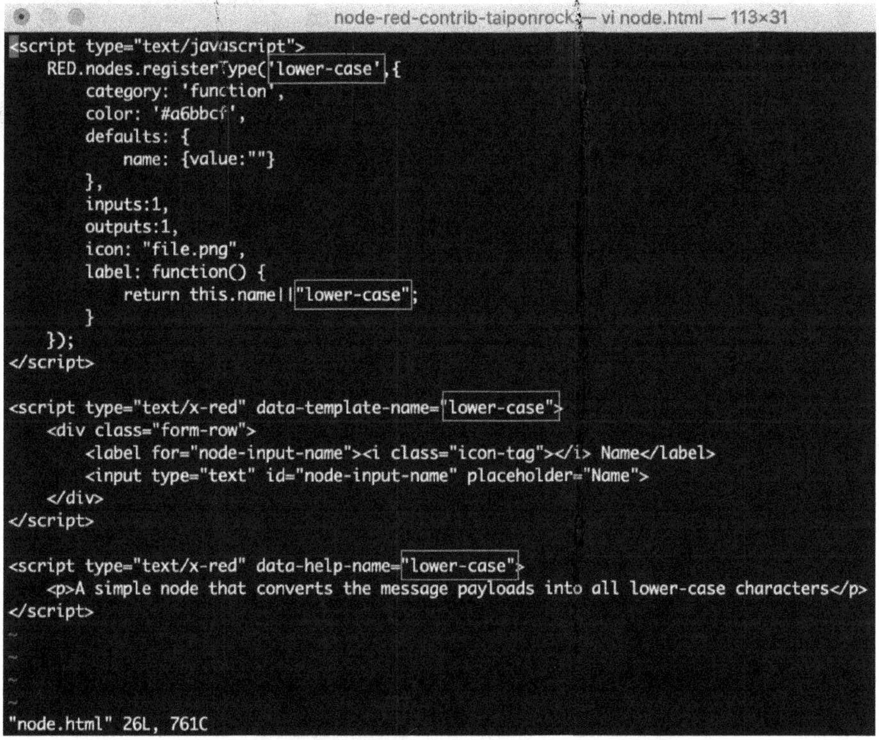
node-red-contrib-taiponrock — vi node.html — 113×31
<script type="text/javascript">
RED.nodes.registerType('lower-case',{
category: 'function',
color: '#a6bbcf',
defaults: {
name: {value:""}
},
inputs:1,
outputs:1,
icon: "file.png",
label: function() {
return this.name||"lower-case";
}
});
</script>
<script type="text/x-red" data-template-name="lower-case">
<div class="form-row">
<label for="node-input-name"><i class="icon-tag"></i> Name</label>
<input type="text" id="node-input-name" placeholder="Name">
</div>
</script>
<script type="text/x-red" data-help-name="lower-case">
<p>A simple node that converts the message payloads into all lower-case characters</p>
</script>
"node.html" 26L, 761C

Figure 13.13 – Before modifying node.html

This is how the `node.html` file looks after being modified:

```
node-red-contrib-taiponrock — vi node.html — 113×31
<script type="text/javascript">
    RED.nodes.registerType('taiponrock',{
        category: 'function',
        color: '#a6bbcf',
        defaults: {
            name: {value:""}
        },
        inputs:1,
        outputs:1,
        icon: "file.png",
        label: function() {
            return this.name||"taiponrock";
        }
    });
</script>

<script type="text/x-red" data-template-name="taiponrock">
    <div class="form-row">
        <label for="node-input-name"><i class="icon-tag"></i> Name</label>
        <input type="text" id="node-input-name" placeholder="Name">
    </div>
</script>

<script type="text/x-red" data-help-name="taiponrock">
    <p>A simple node that converts the message payloads into all lower-case characters</p>
</script>
~
~
~
~
```

Figure 13.14 – After modifying node.html

After restarting Node-RED, you can see that it has been renamed correctly:

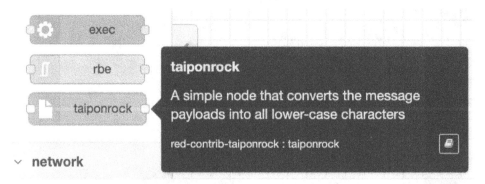

Figure 13.15 – Your node has been renamed

Next, we will see how we can change the code of a particular node.

Changing the node code

The main parts that implement node processing are as follows:

1. Change the code. You can change the processing of the node by rewriting `msg.payload = msg.payload.toLowerCase ();` defined in this part of `node.js`:

```
(abridgement)
node.on ('input', function (msg) {
        msg.payload = msg.payload.toLowerCase ();
        node.send (msg);
}
(abridgement)
```

Here, to make the work easier to understand, let's change to a method that only returns the character string of your name or nickname.

2. Let's rewrite `node.js` as follows:

```
(abridgement)
node.on ('input', function (msg) {
        msg.payload = "Taiponrock";
        node.send (msg);
}
(abridgement)
```

3. Execute the flow.

    Now let's see if it has changed. Use the flow you created earlier. The **lower case** node in this flow has been changed to a node whose name and processing has been changed, but it needs to be redeployed and raised. To make it easier to understand, delete the node that was once the original **lower case** node and relocate it.

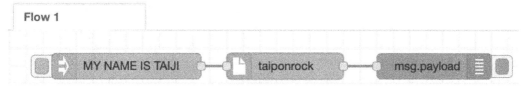

Figure 13.16 – Replace the node you created with the renamed node and execute it

4. Check the result. When you deploy the created flow and execute the **inject** node, you can see that the character string (name or nickname) that was set as a constant in this *Changing the node code* section is displayed in the **debug** tab.

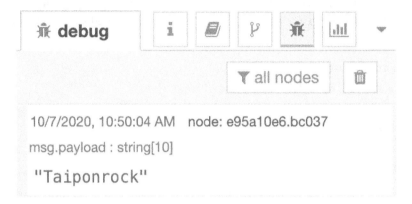

Figure 13.17 – Result of this flow

In the next section, we will see some other node customizing options that we can use.

Other customizing options

In addition to the node name, you can customize your own node in a lot of different ways, such as node color, node icon, node category, node function, and so on. For details, please see this official document: `https://nodered.org/docs/creating-nodes/appearance`.

Now that we have tested and customized the node in the local environment, let's publish the node in the Node-RED library.

Publishing your own node as a module in the Node-RED Library

Here, we will publish the created node in the Node-RED library. To do that, some work is required. So far, you have created your own node and confirmed that it can be used only in your environment. However, since it is a unique node created by you, let's publish it on the internet and have everyone in the world use it. To do this, you need to publish your own node to a location called the Node-RED library, which can be found here: `https://flows.nodered.org/`.

> **Important note**
>
> The Node-RED library is a community-based place to publish nodes and flows. Therefore, you should avoid exposing incomplete or useless nodes. This is because the Node-RED users should be able to find the nodes that they want, and it is not desirable to have a mix of unwanted nodes.
>
> So, although this chapter will explain how to publish nodes, please avoid exposing test nodes or sample node-level ones.

Publishing the node you created

Follow these steps to publish your own node in the Node-RED library:

1. Maintain a README.md file.

    We will write the node description in the README.md file. English is the best language to write in, considering that English is a universal language.

    For example, it is desirable to describe the following contents:

- Overview explanation

- How to use the node

- Screenshot

- Sample flow using this node

- Prerequisite environment

- Change log

    In this section, since it is a hands-on tutorial, only the outline and usage will be written in the README.md file. Please update README.md with the following contents:

```
# node-red-contrib-<Any specified string>


## Overview
This node is a node for forcibly converting all the
alphabet character strings passed as input to the
character string "Taipon rock".

Even if the input parameter passed is not a character
string, "Taiponrock" is forcibly returned.

In this process, it is a wonderful node that changed the
```

```
sample node that was executing toLowerCase, which is an
instance method of String object in JavaScript, to a
process that just returns a meaningless constant.
```

```
## how to use
It is used when you want to forcibly convert all the
character strings of the parameters to be passed to
"Taiponrock".
```

2. Upload files – make sure you have five files: node.js, node.html, package.
    json, README.md, and LICENSE in the directory (it doesn't matter if package.
    lock.json is included):

Figure 13.18 – Check these five files

Upload these files to the repository on GitHub. You should have done the work
in the cloned repository directory, but if you are in another location, move to that
repository directory. Then, execute the following command:

```
$ git add .
$ git commit -m "Node has been published"
$ git push
```

When the push finishes without error, you can see that the target file has been uploaded
in the repository on GitHub:

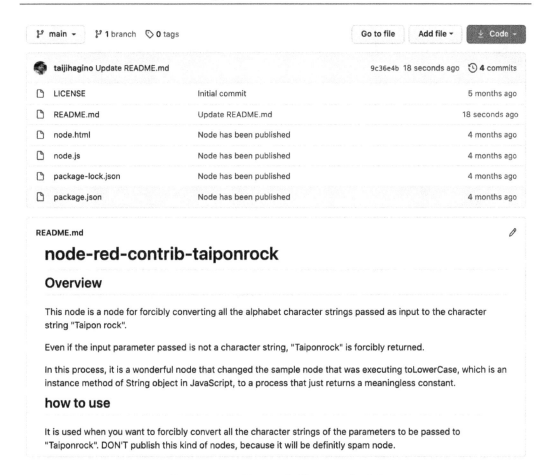

Figure 13.19 – Your node files are uploaded

3. Publish your node (npm publish).

    Now let's expose the node as a module. Upload the set of files using the npm command. Again, work in the cloned repository directory:

    ```
    $ npm adduser
    $ npm publish
    ```

    You will be asked to confirm the version when you run npm publish. Don't forget to edit package.json to increase the version number, as the version must be up when you run npm publish a second time or later.

When publish is completed normally, it will be published at https://www.npmjs.com/package/node-red-contrib-<arbitrary character string>.

An example is https://www.npmjs.com/package/node-red-contrib-taiponrock:

Figure 13.20 – Your node has been published on npm

4. Register the created node from **Adding a node** of the Node-RED library.

5. In **Add your node to the Flow Library**, enter the name of the node you created and click the **add node** button:

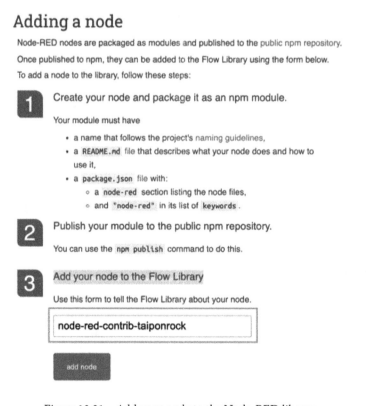

Figure 13.21 – Add your node to the Node-RED library

When the registration is complete, you can see that the created node has been added to the library:

Figure 13.22 – Your node has been published in the Node-RED library

It takes about 15 minutes for the registration of a new node. Please note that the node you registered via the Node-RED flow editor cannot be found without complete registration on the Node-RED library.

If you upgrade the version and publish it again, please refresh from your node's page of the Node-RED Library and click **check for update** in the **Actions** panel on the right side of the node screen:

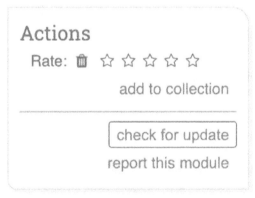

Figure 13.23 – Check for the update of your node's status

Next, let's see how to delete the node published by you.

Deleting the node you published

Be careful when deleting published nodes. Currently (as of October 2020), according to npm's package unpublish policy, the unpublish deadline is within 24 to 72 hours of publication. In addition, it is possible to unpublish packages that have little effect on specific conditions, such as less than 300 downloads even for 72 hours or more.

This information is expected to be updated from time to time, so please refer to the npm official website for the latest information: `https://www.npmjs.com/policies/unpublish`.

After unpublishing, please refresh from your node's page of the Node-RED library in the same way as when updating. Click the request refresh at the bottom of the **Actions** panel on the right side of the node screen.

To unpublish, execute the following command in the module directory (the directory of the cloned repository):

```
$ npm unpublish --force node-red-contrib- <arbitrary string>
```

If this command completes successfully, the module unpublishing is successful.

Installing the node you published

It is recommended that you wait at least 15 minutes after completing adding your node to the **Node-RED Library**.

In Node-RED of the local environment, I reflected the self-made node so that it can be used as it is. I also published it to npm for publication and registered the node in the Node-RED library. Anyone should now be able to use this node.

Here, let's try and check whether the node created this time can be installed and used without any problems from Node-RED of IBM Cloud. Please follow these steps:

1. Log into IBM Cloud, create a Node-RED service, and launch the Node-RED flow editor.

2. Open **Manage Palettes** in the flow editor:

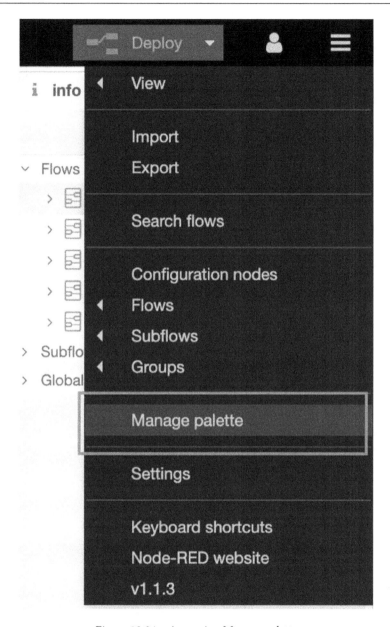

Figure 13.24 – Access ing Manage palette

3. Select the **Install** tab and start typing the name of your node you created in the search field.

    If the node you created is displayed in the search results, it means that it is open to the public and is the target of installation.

4.   Click the **Install** button to install.

If it is not displayed in the search results, you must have not waited for 15 minutes after node registration. Please try again after 30 minutes or 1 hour. If you still do not find your node, there may be some other cause, so please review the procedure you have done so far and try again:

Figure 13.25 – Search for and install your node

5.   Confirm that the node you created on the palette is installed, create a flow as shown in the following figure, and execute the **inject** node:

Figure 13.26 – Make the flow

In the example, the self-made node is inserted between the flows prepared by default when the Node-RED flow editor is started for the first time.

6.   After running the **inject** node, verify that the results are displayed in the **debug** window:

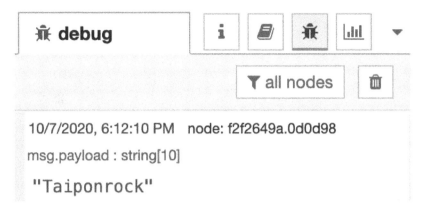

Figure 13.27 – Result of this flow

Great job! You now know how to make your own node and publish it.

Summary

Congrats! In this chapter, you learned how to create your own node, how to customize it, and how to set it from the Node-RED library or your local machine. Creating your own node wasn't as difficult as you might think. If you create the processing content and arrange the appearance based on this procedure, you can publish your own useful node that does not already exist and have developers all over the world use it!

Also, at the end of this book, I'll give you a brief introduction to the Node-RED user community, so be sure to check that out as well.

Appendix
Node-RED User Community

Node-RED is still evolving as an open source tool. Behind the scenes, not only the creators of Node-RED but also its many users are a big force in shaping the tool and contributing to it.

I believe that the user market of Node-RED will grow even more in the future. Here, we will introduce the user community; please do actively participate in the user community, whether you are just starting to use Node-RED or have already been using it for some time.

Node-RED Community Slack

In Node-RED's Slack, creators and users talk about various topics. You can also give feedback to the Node-RED core team.

In addition, the number of channels that support local languages is gradually increasing, so anyone can easily enjoy the conversation: `https://nodered.org/slack/`

Node-RED Forum

The Node-RED Forum gives you support from users or creators on technical issues and development topics. You will get more benefit from it when used in conjunction with the Slack channel mentioned previously: `https://discourse.nodered.org/`

Japan User Group

This is a Node-RED user community for Japan, which Taiji, the author of this book, belongs to and organizes. Its representative is Atsushi Kojo of Uhuru. The information provided is mainly in Japanese, but recently the number of participants from outside Japan have increased, and communication in English can also be found. Once a year, we also have a global Node-RED conference called Node-RED Con: `https://nodered.jp/`

`Packt.com`

Subscribe to our online digital library for full access to over 7,000 books and videos, as well as industry leading tools to help you plan your personal development and advance your career. For more information, please visit our website.

Why subscribe?

- Spend less time learning and more time coding with practical eBooks and Videos from over 4,000 industry professionals

- Improve your learning with Skill Plans built especially for you

- Get a free eBook or video every month

- Fully searchable for easy access to vital information

- Copy and paste, print, and bookmark content

Did you know that Packt offers eBook versions of every book published, with PDF and ePub files available? You can upgrade to the eBook version at `packt.com` and as a print book customer, you are entitled to a discount on the eBook copy. Get in touch with us at `customercare@packtpub.com` for more details.

At `www.packt.com`, you can also read a collection of free technical articles, sign up for a range of free newsletters, and receive exclusive discounts and offers on Packt books and eBooks.

Other Books You May Enjoy

If you enjoyed this book, you may be interested in these other books by Packt:

Building Low-Code Applications with Mendix

Bryan Kenneweg , Imran Kasam , Micah McMullen

ISBN: 978-1-80020-142-2

- Gain a clear understanding of what low-code development is and the factors driving its adoption

- Become familiar with the various features of Mendix for rapid application development

- Discover concrete use cases of Studio Pro

- Build a fully functioning web application that meets your business requirements

- Get to grips with Mendix fundamentals to prepare for the Mendix certification exam

- Understand the key concepts of app development such as data management, APIs, troubleshooting, and debugging

Practical Python Programming for IoT

Gary Smart

ISBN: 978-1-83898-246-1

- Understand electronic interfacing with Raspberry Pi from scratch
- Gain knowledge of building sensor and actuator electronic circuits
- Structure your code in Python using Async IO, pub/sub models, and more
- Automate real-world IoT projects using sensor and actuator integration
- Integrate electronics with ThingSpeak and IFTTT to enable automation
- Build and use RESTful APIs, WebSockets, and MQTT with sensors and actuators
- Set up a Raspberry Pi and Python development environment for IoT projects

Packt is searching for authors like you

If you're interested in becoming an author for Packt, please visit authors. packtpub.com and apply today. We have worked with thousands of developers and tech professionals, just like you, to help them share their insight with the global tech community. You can make a general application, apply for a specific hot topic that we are recruiting an author for, or submit your own idea.

Leave a review - let other readers know what you think

Please share your thoughts on this book with others by leaving a review on the site that you bought it from. If you purchased the book from Amazon, please leave us an honest review on this book's Amazon page. This is vital so that other potential readers can see and use your unbiased opinion to make purchasing decisions, we can understand what our customers think about our products, and our authors can see your feedback on the title that they have worked with Packt to create. It will only take a few minutes of your time, but is valuable to other potential customers, our authors, and Packt. Thank you!

Index

I

J

L

M

N